BIOCHEMISTRY PRIMER FOR EXERCISE SCIENCE

Michael E. Houston, PhD
University of Waterloo
Waterloo, Ontario, Canada

Human Kinetics

Library of Congress Cataloging-in-Publication Data

Houston, Michael E., 1941-
Biochemistry primer for exercise science / Michael E. Houston.
p. cm.
Includes bibliographical references and index.
ISBN 0-87322-577-5
1. Biochemistry. 2. Exercise--Physiological aspects. I. Title.
QP514.2.H68 1995
574.19'2--dc20 94-41211
CIP

ISBN: 0-87322-577-5

Acquisitions Editor: Richard A. Washburn, PhD; **Developmental Editors:** Mary E. Fowler and Marni Basic; **Assistant Editors:** Julie Ohnemus, Bill Hechler, and Henry Woolsey; **Copyeditor:** Frances Purifoy; **Proofreader:** Kathy Bennett; **Indexer:** Theresa J. Schaefer; **Typesetter:** Bob Chapdu; **Text Layout:** Bob Chapdu and Francine Hamerski; **Text Designer:** Jody Boles; **Cover Designer:** Thomas Bradley; **Illustrator:** Studio 2D; **Printer:** United Graphics.

Printed in the United States of America 10 9 8 7 6 5 4 3 2 1

Human Kinetics
P.O. Box 5076, Champaign, IL 61825-5076
1-800-747-4457

Canada: Human Kinetics, Box 24040, Windsor, ON N8Y 4Y9
1-800-465-7301 (in Canada only)

Europe: Human Kinetics, P.O. Box IW14, Leeds LS16 6TR, United Kingdom
(44) 1132 781708

Australia: Human Kinetics, 2 Ingrid Street, Clapham 5062, South Australia
(08) 371 3755

New Zealand: Human Kinetics, P.O. Box 105-231, Auckland 1
(09) 523 3462

To Cheri, my best friend, who has shown me what dedication really means.

CONTENTS

PREFACE

Curricula in kinesiology, exercise science, physical education, and similar departments have changed dramatically in recent years. One notable example is the inclusion of biochemistry as a required course. An understanding of biochemistry opens new doors to the undergraduate student, not only because exciting developments are taking place in this field but also because it enhances understanding of the human organism during physical activity. Graduate students and even faculty are demonstrating a fresh interest in biochemistry because they want to understand the molecular events dominating the field of exercise science.

For the past 25 years, the University of Waterloo has required a course in biochemistry for both kinesiology and health studies students. This course has undergone constant evolution but has consistently focused on the biochemistry of humans in relation to physical activity and nutrition. Unfortunately, one of the problems for students studying human biochemistry is access to a textbook that can supplement lecture material. Most biochemistry textbooks do not concentrate on humans, and those that do often contain more than 1,000 pages, yet omit reference to exercise and sport.

This book is devoted to students with a primary interest in exercise physiology, and a need for molecular fundamentals. It has been tested with more than 700 students over four years and has been shown to be most relevant to students simultaneously studying exercise physiology. Students who have taken a semester of general biology and general chemistry, including some organic chemistry, should have no difficulty with the subject material. The appendix provides a brief review of relevant background material for those needing a quick brush up or more serious study for those with a gap between current knowledge and their new interest in biochemistry.

The course emphasizes basic elements of carbohydrate, fat, and protein metabolism. A brief yet comprehensive section on molecular biology highlights the major players in this subject and addresses simple concepts of transcription regulation. Students completing this course usually have no difficulty understanding what is happening at the research fronts in exercise physiology. A background in biochemistry also makes the science of nutrition more interesting and relevant. Finally, a fundamental understanding of biochemistry enhances one's ability to function as an educated person in our increasingly complex society.

Acknowledgments

I would like to thank Bengt Saltin, a critical thinker, colleague, and friend, who inspired me more than he knows. I wish to thank Jim Stull, who gave me the opportunity to work and learn at Southwestern Medical Center, and Robert W. McGilvery, a writer whom I never met, who showed me that biochemistry can be fun. Finally, to more than 3,000 Kinesiology 317 students, who may have taken biochemistry reluctantly but who inspired me to strive for the perfect means of communication.

The Major Players

Everything you do, everything you are, and everything you become depends on the actions of thousands of different proteins. These are large molecules composed of individual units called amino acids. Twenty different amino acids are used to make proteins, as few as 100 amino acids form a small protein. The order in which they are arranged in a specific protein is critical. This information is stored in the nucleus in discrete elements called genes, which are sections of the DNA molecules that make up our chromosomes. Four different bases comprise the information component of DNA. The sequence of these bases in DNA molecules spells out the sequence of amino acids in proteins. In Part I, we will look at amino acid building blocks, their properties as acids and bases, how they are assembled to make proteins, and a description of protein structure. Chapter 2 focuses on the body's major protein action molecules, the enzymes. This section concludes with the structure of the master blueprint—DNA and its working partner RNA.

Chapter 1

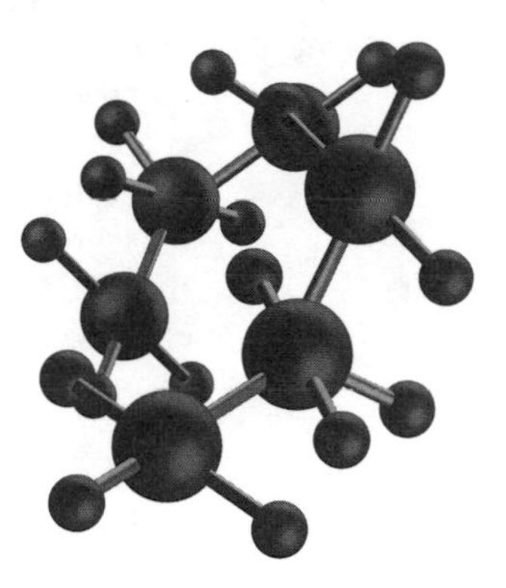

Amino Acids, Peptides, and Proteins

Proteins are the molecules responsible for what happens in cells and organisms. We can consider them the action molecules because they can be enzymes that catalyze all the chemical reactions in the body; receptor molecules inside cells, in membranes, or on membranes that bind specific substances; contractile proteins involved in the contraction of skeletal, smooth, and heart muscle; or transport molecules that move substances in the blood, within cells, and across membranes. Proteins are also parts of bones, ligaments, and tendons, and in the form of antibodies or receptors on lymphocytes they help protect us from disease.

Twenty different amino acids are used to make proteins. The genes in our cell nuclei contain the information needed to specify what amino acids are used in making a protein and in what order. The genes in chromosomes are segments of DNA, large molecules containing four different bases. The sequence of bases in a gene spells out the sequence of amino acids for a protein. The base sequence in a gene is copied by making a new molecule, messenger ribonucleic acid (mRNA), in a process known as transcription. The mRNA information is then used to order the binding of the 100 or more specific amino acids to make a protein. This step, known as translation, takes place outside of the nucleus. Proteins are continually being turned over: Old protein molecules are broken down to their constituent amino acids in a process known as degradation, and new protein molecules are synthesized. This continual synthesis and degradation of proteins in cells is known as turnover (Figure 1.1). We can express the rate of turnover of an individual protein as the time taken for one half of the protein molecules to be

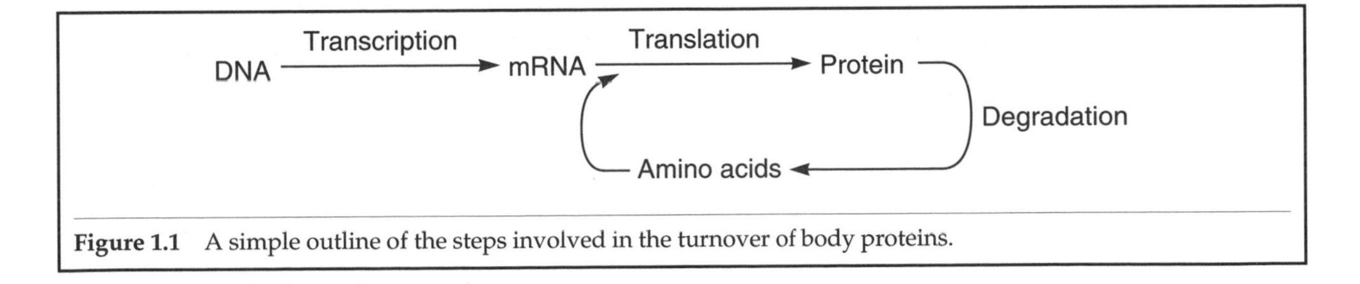

Figure 1.1 A simple outline of the steps involved in the turnover of body proteins.

replaced, that is, the half-life. Some proteins have a short half-life, measured in minutes or hours, while others have a half-life that may be expressed in weeks. An interesting research area involves the role of exercise training on the turnover of skeletal muscle proteins.

The Nature of Amino Acids

Figure 1.2 illustrates what is typically shown for the general structure of an amino acid. In this figure, each amino acid has a carboxyl group, an amino group, a central carbon atom identified as the alpha (α) carbon (carbon 2), and a variable part known as its side chain indicated by the letter *R*. In fact, 19 of the 20 common amino acids have this general structure. What makes each amino acid unique is its side chain or R group. Each amino acid has (a) a specific side chain (R group), (b) a name, (c) a three-letter short form, and (d) a single capital letter to represent it. Differences in the properties of the 20 common amino acids are based on their side chains. Figure 1.3 shows examples of four common amino acids, including the name, three-letter short form and single letter identification, and the side chain.

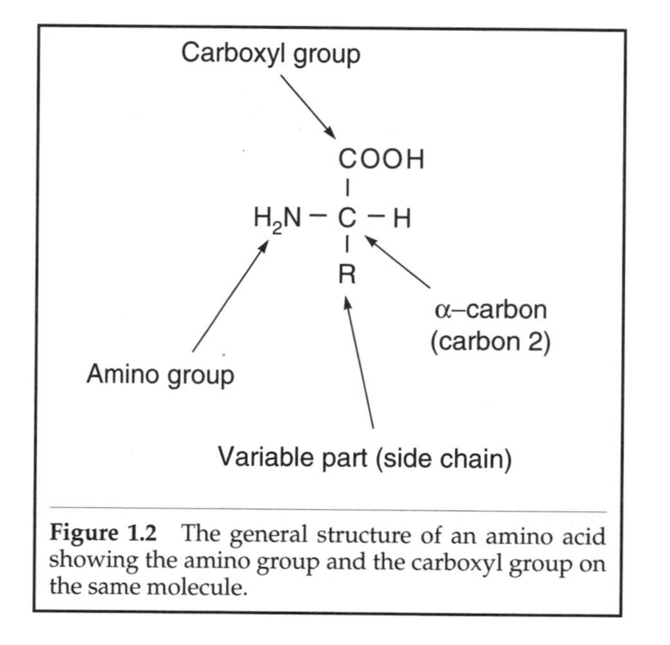

Figure 1.2 The general structure of an amino acid showing the amino group and the carboxyl group on the same molecule.

The general amino acid structure in Figure 1.2 never occurs exactly as shown because of the acid-base properties of the amino and carboxyl groups. Figure 1.4 shows the general structure of an amino acid at pH 7.

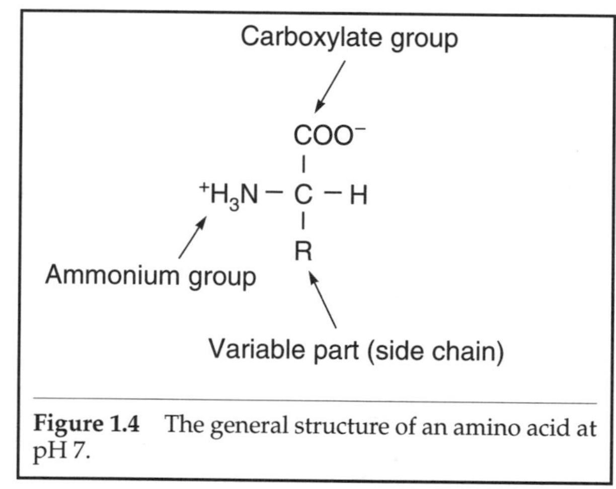

Figure 1.4 The general structure of an amino acid at pH 7.

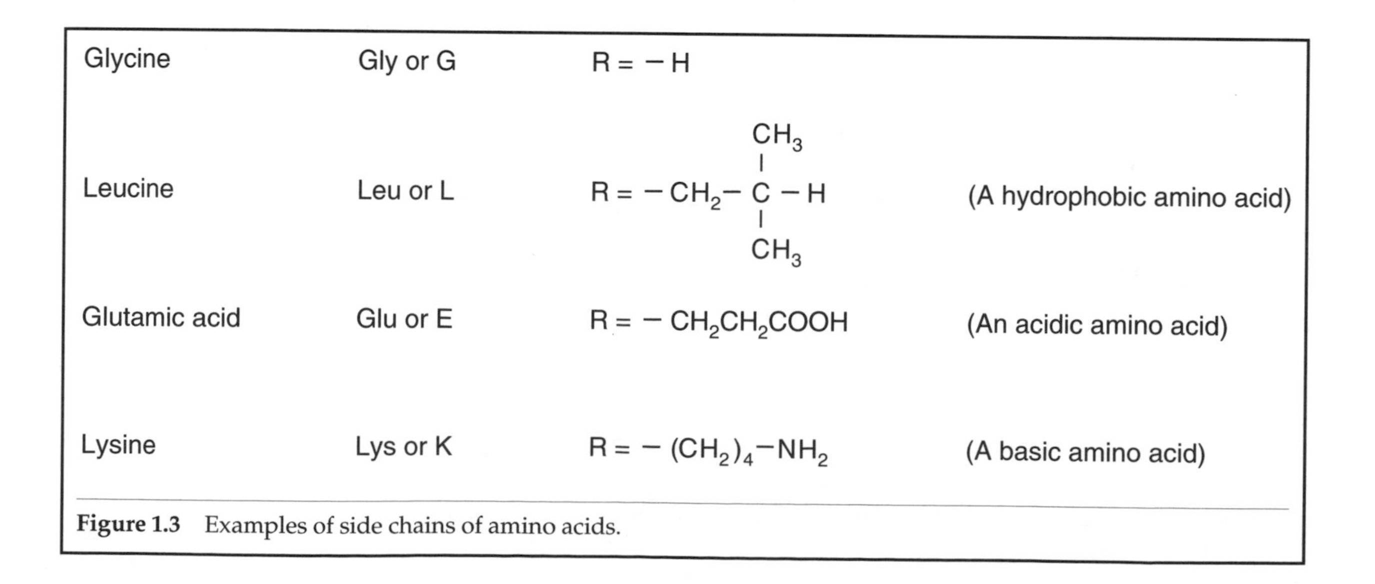

Figure 1.3 Examples of side chains of amino acids.

Acid-Base Properties of Amino Acids

The dissociation of the general weak acid (HA) is written as

$$\underset{\text{acid}}{HA} \longleftrightarrow \underset{\text{proton}}{H^+} + \underset{\text{conjugate base}}{A^-}$$

Because weak acids do not dissociate completely, we can write an equilibrium expression for the reversible dissociation using square brackets to represent concentration

$$\frac{[H^+]\,[A^-]}{[HA]} = K_a$$

The stronger the acid (HA), the more it is dissociated, the more the equilibrium is shifted to the right, and the larger the value for K_a. Therefore, the larger the numerical value for K_a, the stronger the acid.

Suppose we dissolve HA in a solution containing the sodium salt of A^- (i.e., NaA) such that at equilibrium, the concentration of undissociated HA is equal to the concentration of A^-. Now the above equilibrium equation is simplified to

$$[H^+] = K_a$$

because the other terms ([HA] and $[A^-]$) cancel out.

Now we will take the negative logarithm of both sides of this equation; i.e.,

$$-\log\,[H^+] = -\log\,K_a$$

$$\text{or } pH = pK_a$$

By definition, pH is the negative logarithm of $[H^+]$. Thus pK_a is the negative logarithm of K_a.

The pK_a for an acid is the pH of a solution where the acid is one-half dissociated, i.e., 50% of the molecules are dissociated and 50% are not dissociated. Therefore, the smaller the value of pK_a for an acid, the stronger the acid.

All amino acids have at least one acid group (proton donor) and one basic group (proton acceptor). Amino acids may have both acid and base properties; that is, they are amphoteric.

In amino acids, the major acid groups are the carboxyl (−COOH) and the ammonium ($-NH_3^+$) groups. These protonated forms can each give up a proton. The major base groups are the carboxylate ($-COO^-$) and amino ($-NH_2$) groups, which are unprotonated and can accept a proton.

Now let us look at equations describing the ionization of the protonated (acid) forms of the groups.

$$-COOH \longleftrightarrow -COO^- + H^+$$

will have a K_a value and hence a pK_a.

$$-NH_3^+ \longleftrightarrow -NH_2 + H^+$$

will also have a K_a and thus a pK_a.

The carboxyl group (−COOH) is a stronger acid than ammonium ($-NH_3^+$) and will have a lower pK_a value (or its K_a value will be larger). Conversely, the amino group will be a better proton acceptor (base) than the $-COO^-$ and will have a larger pK_a.

The Henderson-Hasselbalch equation describes the relationship between the pH of a solution, the pK_a of the weak acid group, and the concentration ratio of the acid (HA) and its conjugate base (A^-).

$$pH = pK_a + \log\frac{[\text{conjugate base}]}{[\text{acid}]}$$

This equation predicts the main form of an ionizable group in a solution at a known pH if we know the pK_a for the ionizable group. Let us carry out a mental experiment using the amino acid alanine. We dissolve alanine in a beaker containing HCl (hydrochloric acid) such that the pH of the solution is 1.0. Figure 1.5 shows alanine at a pH of 1.0. Note that the pK_a values for the two ionizable groups are included and described as pK_1 and pK_2.

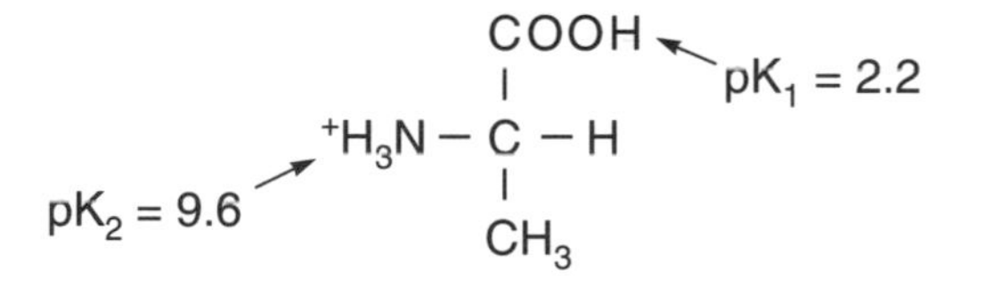

COOH ← $pK_1 = 2.2$
$^+H_3N - C - H$
$pK_2 = 9.6$ →
CH_3

Figure 1.5 The structure of the amino acid alanine in a solution of pH1.0. The pK_a values for the two dissociable groups are shown as pK_1 and pK_2.

At a pH of 1.0, both ionizable groups are protonated because, from the Henderson-Hasselbalch equation, when the pH is less than the pK_a, the ratio of conjugate base to acid is less than one. In other words, the ionizable group will be predominantly protonated and in a solution with a pH of 1.0, it will have a net charge of +1.

Let us now add, drop by drop, a solution of sodium hydroxide (NaOH) to our solution of alanine (which, before adding the NaOH, has a pH of 1.0). The addition of NaOH will increase the pH of the solution, but what will it do to our ionizable groups? From your basic understanding of chemistry, you should see that the added OH^- will start removing a proton from the ionizable groups to form water. However, will the proton first be removed from the −COOH or from the $-NH_3^+$? It will be removed first from the group that is the best acid (that is, the one having the lower

pK_a value). This will be $-COOH$. If we carefully monitor the pH of our alanine solution as we add the NaOH, the pH will not change much because the OH^- we add is mainly removing a proton from the $-COOH$ group.

The structure of alanine after we have added enough NaOH to make the solution pH equal to 2.2 will be an equal mixture of carboxylate and carboxylic acid forms of the alanine molecules because a pH of 2.2 equals the pK_a for this group, described as pK_1. Remember, pK_a refers to the pH when the acid group is 50% dissociated. However, nothing happens to the ammonium group of alanine at this pH because its pK_a (i.e., pK_2) is much higher at 9.6. As we continue to increase the pH by adding more NaOH, we remove all protons from the remaining undissociated $-COOH$, and the pH increases more. If we stop adding NaOH when our pH meter reads 5.9, the structure of alanine will look like that shown in Figure 1.6. This structure shows that alanine has no net charge but it has one positive and one negative charge. This is known as the Zwitterion form for alanine. The pH where a molecule has an equal number of positive and negative charges (and therefore no net charge) is its isoelectric point—designated pI. Notice that 5.9 is one half the value of the sum of the pK_1 and pK_2 for alanine.

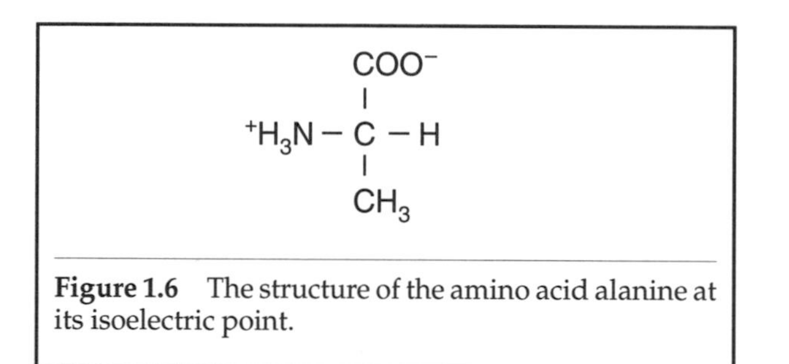

Figure 1.6 The structure of the amino acid alanine at its isoelectric point.

If we continue to add NaOH to our solution of alanine until the pH increases beyond 5.9, we start removing the proton from the ammonium ($-NH_3^+$) group. Thus, at a pH greater than the pK_2 for alanine, the predominant structure will look like that in Figure 1.7. Notice that this structure (found at a pH greater than 9.6) reveals that alanine will have a net negative charge. At a pH greater than 9.6, alanine's two dissociable groups will be in their base form. The amino group and the carboxylate group can act as bases because they both can accept protons. If we add acid to a solution whose pH is greater than 9.6, the amino group will be the first to accept protons. After the amino group is converted to ammonium, protons will be added to the carboxylate group.

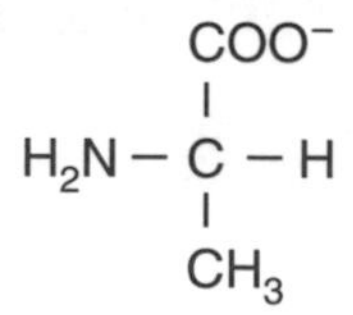

COO^-
$H_2N - C - H$
CH_3

Figure 1.7 The predominant structure of alanine at a pH greater than 10.

Monoamino dicarboxylic amino acids, such as aspartic acid (see Figure 1.8), have two groups that can be carboxyl or carboxylate and only one group that can be amino or ammonium. We call such amino acids acidic amino acids. Figure 1.8 shows the structure of aspartic acid when in solution at a pH equal to its isoelectric point. Note that it has one positive charge, one negative charge, and no net charge. This is exactly in agreement with the definition of isoelectric point. However, the actual value for the pI for aspartic acid is quite low because it must be greater than pK_1 and less than pK_2, or exactly one-half their sum (i.e., 2.8). The only group that can have a positive charge is the ammonium group. However, there are two carboxyl groups which can have a negative charge. The one that has the negative charge will have the lower pK_a value because it is the best proton donor.

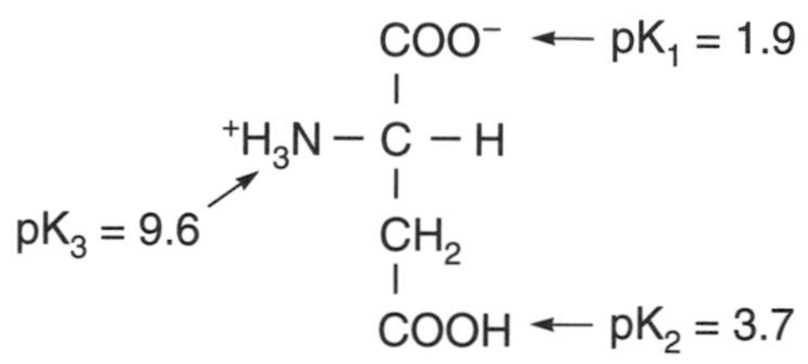

COO^- ← $pK_1 = 1.9$
$^+H_3N - C - H$
$pK_3 = 9.6$
CH_2
$COOH$ ← $pK_2 = 3.7$

Figure 1.8 The structure of aspartic acid at its isoelectric point. The pK_a values for the three dissociable groups are shown in order of increasing magnitude.

Diamino monocarboxylic amino acids, such as lysine (see Figure 1.9), have two amino or similar type groups and only one carboxyl group and are often called basic amino acids. Figure 1.9 shows the structure of lysine at its isoelectric point. Notice that lysine has no net charge due to an equal number of positive and negative groups. Determining which of the two amino groups will be in the ammonium form is easy—it will be the one with the higher pK_a value, or the best base or proton acceptor. The pI for lysine will be one half the sum of the pK_a values for the two amino groups, or one half the sum of pK_2 and pK_3.

Figure 1.9 The structure of lysine at its isoelectric point. The pK_a values for the dissociable groups are shown.

The following rules will never fail you in working out acid-base problems with amino acids:

RULE 1: When the pH of an amino acid solution is less than the pK_a of the ionizable group in question, that group will exist primarily in protonated (acid) form.

RULE 2: When the pH of an amino acid solution is greater than the pK_a of the ionizable group in question, that group will exist primarily in unprotonated (base) form.

RULE 3: When the pH of an amino acid solution is equal to the pK_a of the ionizable group in question, one half of the amino acid molecules will have that group in protonated form and the other half will have that group in unprotonated form.

When given an acid-base problem with amino acids or peptides, work out the state (protonated or unprotonated) of each ionizable group, one at a time. The rules always work.

Demonstrate that you understand acid-base properties of amino acids by drawing the predominant structures of aspartic acid and lysine in the blood, which has a pH of 7.4. Figure 1.10 shows these structures. Also draw the structure of the common flavor enhancer monosodium glutamate. Glutamic acid is similar to aspartic acid except that it has an additional methylene (CH_2) in its side chain. Assume that the pK_a values for the three dissociable groups in glutamic acid are the same as those of aspartic acid.

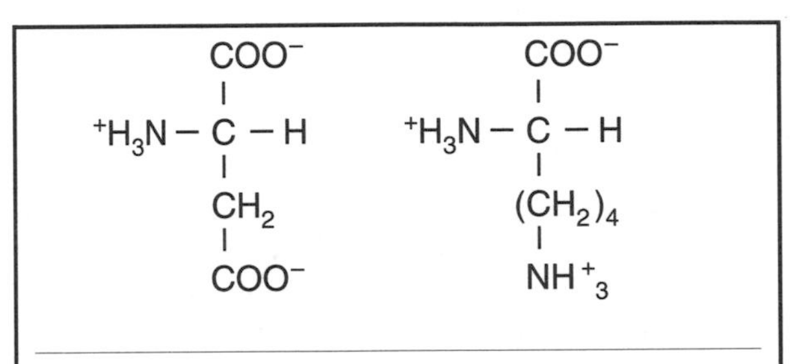

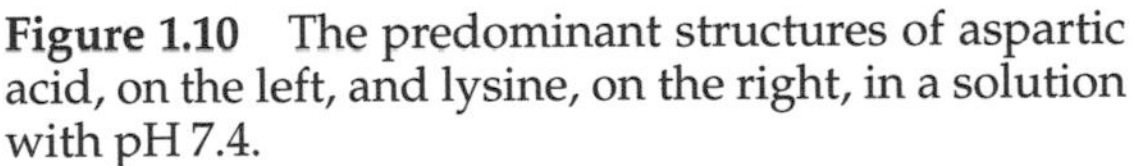

Figure 1.10 The predominant structures of aspartic acid, on the left, and lysine, on the right, in a solution with pH 7.4.

In addition to the major ionizable groups discussed so far (i.e., the alpha and side chain amino and carboxyl groups), others can be quite important (see Figure 1.11). Histidine is often found at the active site of many enzymes, for it influences their catalytic ability.

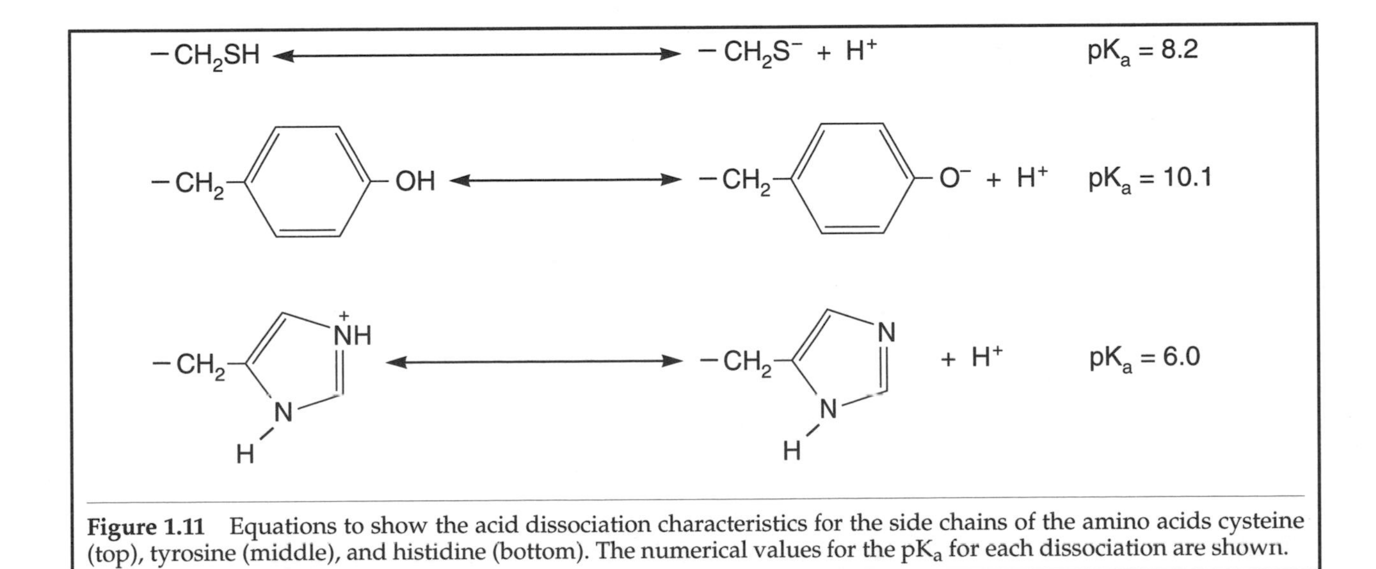

Figure 1.11 Equations to show the acid dissociation characteristics for the side chains of the amino acids cysteine (top), tyrosine (middle), and histidine (bottom). The numerical values for the pK_a for each dissociation are shown.

Stereoisomerism of Amino Acids

All amino acids except glycine have four different groups attached to their alpha carbon. Because of this, the alpha carbon is asymmetric or chiral, with two different ways of arranging these groups; i.e., two different configurations. Figure 1.12 shows the groups around the central (alpha) carbon to be three-dimensional; the dashed lines mean that the bonds are going into the paper, and the wedges mean that the bonds are coming out of the plane of the paper toward you. When the carboxylate group is at the top and going into the paper, the L and D refer to the position of the ammonium group, i.e., on the left side (L) or right side (D or dextro). These stereoisomers (or space isomers) are also enantiomers—pairs of molecules that are nonsuperimposable mirror images of each other. They may be compared to a right and left hand. When you hold a left hand to a mirror, a right hand is the image. Similarly, holding a D-amino acid to a mirror gives an L-amino acid as the image. All naturally occurring amino acids are in the L-configuration.

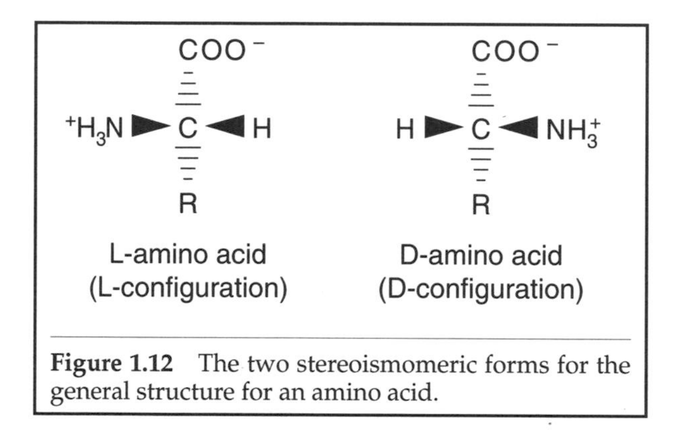

Figure 1.12 The two stereoismomeric forms for the general structure for an amino acid.

The Characteristics of Peptides

Peptides are formed when amino acids join together via their ammonium and carboxylate groups in a specialized form of the amide bond known as a peptide bond. Because the body's amino acids are in an environment with a pH around 7, the amino group is protonated and is an ammonium group, whereas the carboxyl group is unprotonated and is a carboxylate group. Figure 1.13 shows how a peptide is formed from two amino acids. Combination of the ammonium and carboxylate groups yields the peptide bond, and a water molecule is eliminated. The peptide bond is rigid and planar. Peptides are drawn by convention starting with the free alpha ammonium group and ending with the free alpha carboxylate group.

The amino acids in a peptide (or protein) are known as residues. The prefixes describing the number of amino acids in a peptide are known as : di—two, tri—three, tetra—four, penta—five, hexa—six, hepta—seven, octa—eight, nona—nine, and deca—ten. Thus a nonapeptide consists of nine amino acid residues. If there are more than 10 amino acids in a peptide it is generally known as a peptide containing, for example, 25 amino acids. The term polypeptide refers to a large peptide that may contain hundreds of amino acid residues. The term protein is generally used to describe a polypeptide containing more than 100 amino acids having a specific function but this is certainly not a hard and fast rule. The primary structure of a peptide (or protein) is the sequence of amino acids starting from the N-terminus. Peptides are also named from the N-terminus. Remember, the blueprint for the sequence of amino acids in a peptide or protein is provided by the base sequence of deoxyribonucleic acid or DNA in a gene.

Many important hormones are peptides. Examples are the pancreatic hormones insulin and glucagon, which control blood glucose concentration. Growth hormone is a peptide hormone secreted from the anterior portion of the pituitary. Some athletes believe that a diet rich in protein will increase the release of growth hormone, making them more muscular, and other people claim that supplementing the diet with specific amino acids will naturally

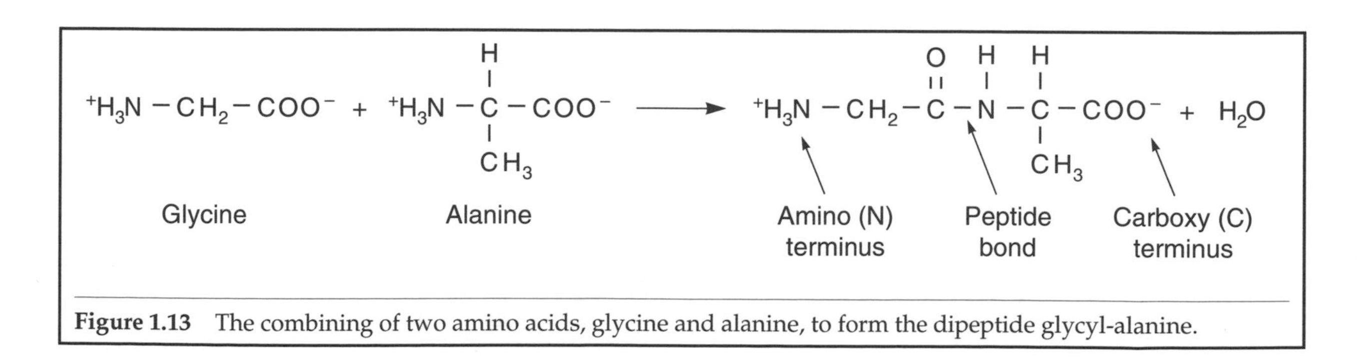

Figure 1.13 The combining of two amino acids, glycine and alanine, to form the dipeptide glycyl-alanine.

stimulate growth hormone release. Unfortunately, the actual results are inconsistent. Other hormones are oxytocin, the endorphins, the hypothalamic releasing hormones, luteinizing hormone, and follicle stimulating hormone. There are also many important peptide growth factors. The number of these is still increasing, with increased attention directed toward cellular growth and its control.

The Structure of Proteins

Proteins are composed of one or more polypeptide chains. Many also contain other substances, such as metal ions (e.g., hemoglobin contains iron); carbohydrates (e.g., glycoproteins contain sugars attached to amino acids—these can be found on cell membranes, pointing into the fluid surrounding cells); and fat or lipid (e.g., blood lipoproteins transport fat such as cholesterol and triglyceride). The precise biological structure of a protein, and hence its function, is determined by the amino acids it contains.

As mentioned, the primary structure of a protein (or peptide) is the sequence of amino acids starting from the N-(amino) terminus. Remember, with 20 different amino acids and at least 100 in a protein, the different combinations are almost unlimited. However, a specific protein in your body must have its proper amino acid sequence. Even replacing one amino acid with another could make a protein totally useless and possibly dangerous. For example, sickle cell anemia, characterized by short-lived erythrocytes of unusual shape, results from the replacement of just one amino acid (valine) with another (glutamic acid) in one chain of the hemoglobin molecule.

Bonds and Interactions Responsible for Protein Structure

Strong, rigid peptide bonds join the individual amino acids together in the protein. Disulfide bonds, illustrated in Figure 1.14, are covalent bonds that join together the side chains of two cysteine residues ($-CH_2SH$) in the same or different polypeptides, resulting in -S-S (disulfide) bonds and the loss of two hydrogen atoms. Disulfide bonds in the same polypeptide chain are intramolecular bonds, whereas those in different polypeptide chains are intermolecular bonds. For example, the polypeptide hormone insulin consists of two peptide chains held together by two intermolecular disulfide bonds.

Hydrogen bonds are weak electrostatic attractions between an electronegative element, such as oxygen or nitrogen, and a hydrogen atom covalently bonded to another electronegative element. Hydrogen bonds between individual water molecules, for example, are responsible for some of the unique properties of water. Figure 1.14 shows the hydrogen bond between the components of a peptide bond, which are important in maintaining overall protein structure. Hydrogen bonds also form between polar parts of some amino acid side chains and those in the protein molecule and water. Although hydrogen bonds are weak, they are important because so many of them are involved in the structure of a protein.

Figure 1.14 shows ionic interactions between oppositely charged groups, as found in proteins, such as the N and C termini and the charged carboxylate or ammonium groups in the side chains.

Amino acids are primarily found in a polar environment because they are in an aqueous medium with charged ions and molecules. Proteins spanning membranes are an exception because the interior of membranes is hydrophobic. Water is a polar molecule and interacts well with polar and charged groups on amino acids. However, 8 of the 20 common amino acids are hydrophobic or nonpolar because their side chains are composed of carbon and hydrogen atoms and thus have no affinity for water. In fact, they can disrupt the relatively organized structure of liquid water. These side chains tend to cluster in the interior of the protein molecule not because they have an affinity for each other but because they cannot interact with water. Fat globules also form tiny spheres in water to keep their exposed surface as small as possible. We call the clustering of hydrophobic groups on amino acids hydrophobic interactions (see Figure 1.14).

Secondary Structure of Proteins

So far, a one-dimensional protein structure has been described; i.e., the sequence of amino acids. However, proteins actually have a three-dimensional structure. Imagine a peptide backbone with the side chains of each amino acid radiating out. The backbone consists of three elements—the alpha carbon, the nitrogen group involved in the peptide bond, and the carbonyl group. The secondary structure is the spatial path taken by these three elements of the peptide backbone. Side chains (R groups) of individual amino acids are not considered in secondary structure. Two common, recognizable structural features of the polypeptide backbone are found in many proteins—the alpha-helix and the beta-sheet, which result from the way the polypeptide backbone organizes itself. They are stabilized by hydrogen bonding between

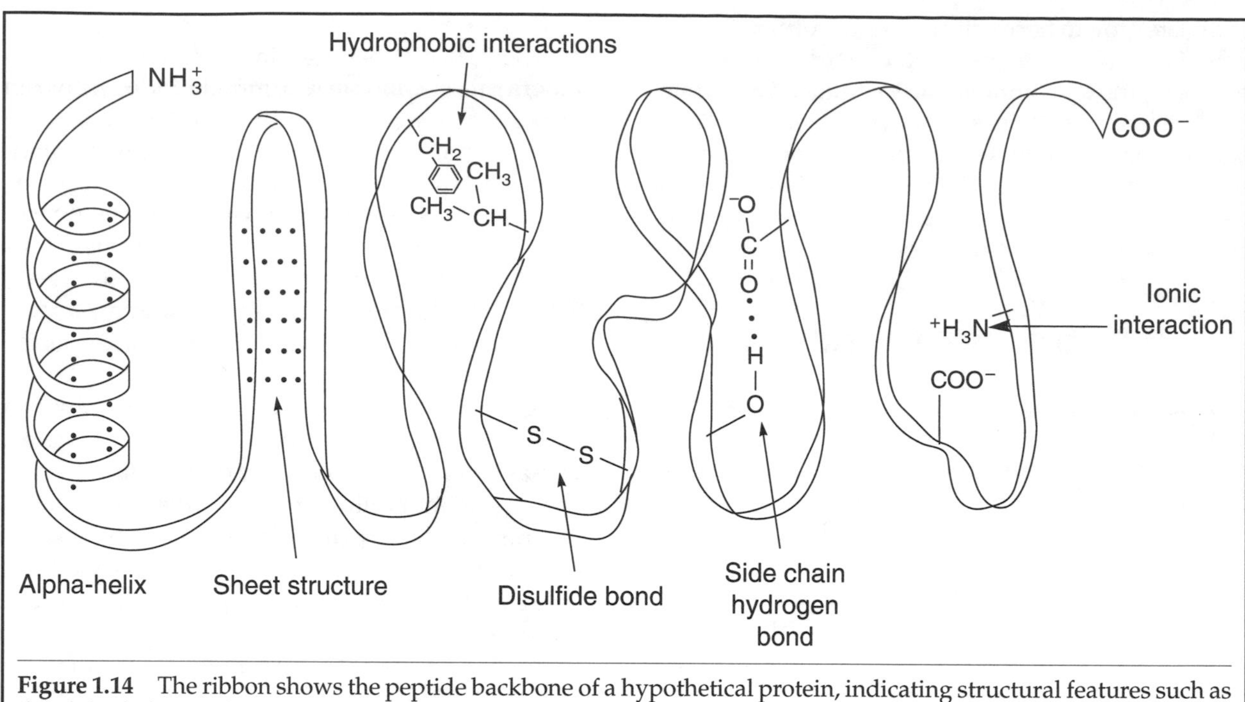

Figure 1.14 The ribbon shows the peptide backbone of a hypothetical protein, indicating structural features such as the alpha-helix and beta-pleated sheet characteristic of protein secondary structure and the hydrogen bonding (dotted line), ionic interactions, hydrophobic interactions, and disulfide bonds, which maintain the three-dimensional conformation of proteins.

neighboring elements of the peptide bond, as shown in Figure 1.14 with dotted lines.

Tertiary Structure of Proteins

The tertiary structure is the overall three-dimensional arrangement of a polypeptide chain, including the secondary structure and any nonordered interactions involving amino acid residues that are far apart. Most proteins are found in aqueous mediums where they assume a compact globular structure maintained by the forces already described. The hydrophobic side chains are buried in the interior of the protein, away from the aqueous medium, with hydrophilic side chains and the N and C termini exposed on the surface where they can interact. The tertiary structure also involves the spatial position of ions or organic groups that are part of the make up of many proteins.

The secondary and tertiary structure of a polypeptide chain depends on the kind and sequence of its amino acids. In the cell, or wherever a protein is found, its overall structure or conformation must be maintained because proteins must recognize and interact with other molecules. However, the conformation of a protein *in vivo* is not fixed but changes in subtle ways as it carries out its particular function. For example, the cross bridges in the contractile protein myosin alter their conformation to generate force during muscle contraction.

Quaternary Structure of Proteins

Many proteins consist of more than one polypeptide chain, each containing its own unique primary, secondary, and tertiary structures. We call these chains subunits. Quaternary structure refers to the arrangement of the individual subunits with respect to each other. The subunits in an oligomeric protein (i.e., one containing more than one subunit) are held together with noncovalent bonds; i.e., hydrogen bonds, electrostatic interactions, and hydrophobic interactions. Oligomeric proteins are common because several subunits allow subtle ways of altering the protein's function. For example, hemoglobin A, the adult form of hemoglobin, is a tetramer consisting of four subunits, two alpha (with 141 amino acids per subunit) and two beta (with 146 amino acids per subunit). Each subunit contains one heme group that binds one oxygen molecule. The quaternary structure of hemoglobin refers to the way the two alpha and two beta subunits interact with each other. The oligomeric nature of hemoglobin enhances its role in loading oxygen in the lung and releasing it at the cell level.

Myosin, a contractile protein, is a hexamer (i.e., consists of six subunits). There are two heavy chains, each with a molecular weight of about 200,000 (often expressed as 200,000 Daltons or 200 kilodaltons—kDa). The two chains wind around each other forming a single long chain for most of their length. At one end, they each form separate globular regions known as heads. In each globular head there are two light-chain polypeptides each with a molecular weight of about 18 to 25 kDa. Two are regulatory light chains and two alkali light chains, which can be extracted with dilute alkali (see Figure 1.15). The globular heads act as the cross bridges that attach to the protein actin to generate force when a muscle contracts. Muscle fiber typing using histochemistry is based on chemical differences in the heavy chains of myosin in different muscle fiber cross sections. Type I fibers have the type I myosin heavy chains, type IIA fibers have IIA heavy chains, and IIB fibers have IIB myosin heavy chains. There are three types of myosin, found in heart muscle (cardiac myosin), smooth muscle, and skeletal muscle. Indeed, myosin is found in all cells.

The term domain refers to part of a protein molecule, usually globular in shape, that has a specific function, such as binding or catalysis. The noncovalent forces that maintain the secondary, tertiary, and quaternary structures of proteins are generally weak. Denaturation refers to a disruption in the overall conformation of a protein with loss of biological activity. When denatured, proteins usually become less soluble. Denaturation can be caused by heat, acids, bases, organic solvents, detergents, agitation (e.g., beating egg whites to produce a meringue), or specific denaturing agents such as urea. If denaturation is not too severe, some proteins may renature, resume their natural conformation, and regain their biological activity.

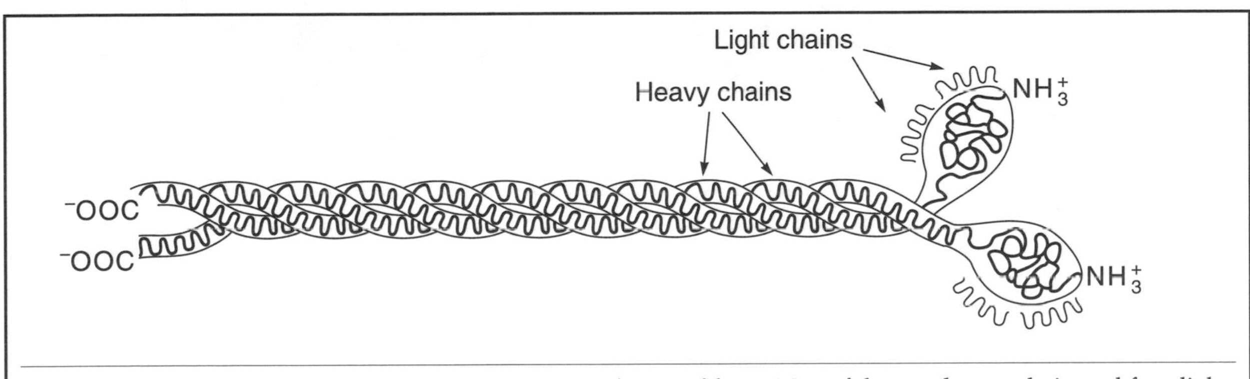

Figure 1.15 Schematic of a myosin molecule showing the possible position of the two heavy-chain and four light-chain subunits.

Summary

Proteins are the action molecules of life, made from 20 different amino acids. Differences among amino acids are based on their characteristic side chains, known as R groups. Amino acids exist in all cells of the body and in the fluids outside the cells. Because each has at least one group that acts as an acid and one that acts as a base, they are said to be amphoteric. All amino acids have at least one charged group, and at the neutral pH associated with life, at least two charged groups. When an amino acid has no net charge, it is said to be a zwitterion. The pH where this occurs is known as the isoelectric point.

Peptides are formed when amino acids are joined together by peptide bonds. In general, proteins are large peptides (polypeptides), usually containing 100 or more amino acids. The primary structure of a peptide or protein is the sequence of amino acid residues starting from the end containing the free amino group, known as the N-terminus. Because of their large size, proteins have higher levels of three-dimensional structure, known as secondary and tertiary structures. If a protein consists of more than one polypeptide, the structural relationship of the polypeptide subunits is described as a quaternary structure. The forces, such as hydrogen bonds and hydrophobic interactions, maintaining the secondary, tertiary, and quaternary structures are generally weak.

CHAPTER 2

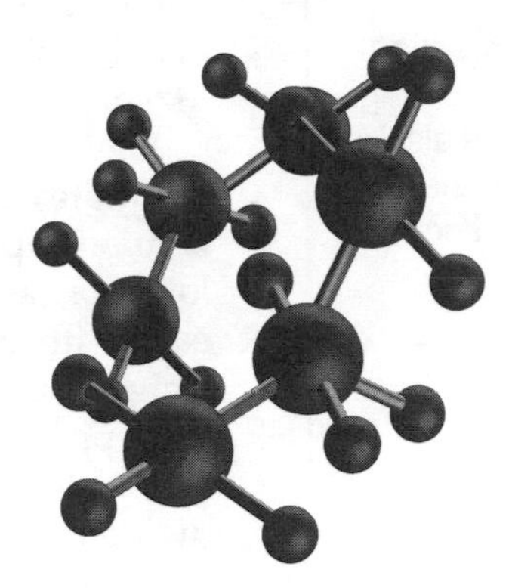

Enzymes

Enzymes are proteins that catalyze the thousands of different chemical reactions in the body. As proteins, they are large molecules, made from the 20 different amino acids. As catalysts, enzymes speed up chemical reactions without being destroyed in the process.

Enzymes as Catalysts

Enzymes speed up chemical reactions by lowering the energy barrier to the reaction—called energy of activation—so that they take place at the low temperature of an organism (378C for a human). Enzymes are highly specific, catalyzing a single reaction or type of reaction. Many enzymes catalyze reactions that, if left by themselves, would eventually reach equilibrium. The enzyme allows the reaction to reach equilibrium faster than if the enzyme were absent but the position of equilibrium remains the same.

The molecule or molecules that an enzyme acts on is known as its substrate or substrates, respectively, which is then converted into a product or products. A part of the large enzyme molecule will reversibly bind to the substrate (or substrates) and then a specific part or parts of the enzyme will catalyze the specific change necessary to change the substrate into a product. The enzyme has amino acid residues that bind the substrate and those that carry out the actual catalysis. There is thus a binding and a catalytic domain, although the term active site often represents both the binding and catalytic domains of the enzyme protein.

We can write a general equation to describe a simple enzyme-catalyzed reaction in which a single substrate (S) is converted into a single product (P). The reaction could be irreversible, with all substrate molecules converted into product molecules, and indicated by an arrow with only one head pointing toward the product. Alternatively, the reaction could be reversible such that given enough time, it establishes an equilibrium, with the ratio of product concentration to substrate concentration a constant described by the equilibrium constant (K_{eq}). Reversible reactions are shown with a double arrow, meaning that the product is also a substrate for the reverse reaction. Figure 2.1 illustrates these two types of reactions and outlines their properties. Most enzyme-catalyzed reactions are considered to take place in discrete steps (see Figure 2.2). Enzyme E first combines reversibly with substrate S to form an initial enzyme-substrate complex ES. This complex is then converted to an enzyme-intermediate complex (EI), which then changes to an enzyme-product complex (EP). EP then dissociates into free product, and the enzyme is released unchanged.

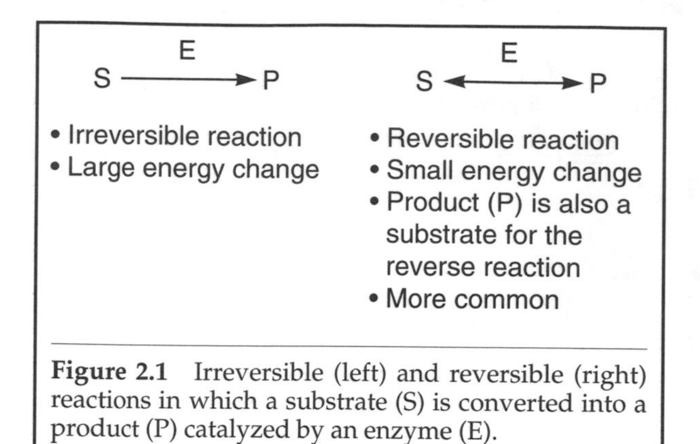

Figure 2.1 Irreversible (left) and reversible (right) reactions in which a substrate (S) is converted into a product (P) catalyzed by an enzyme (E).

Rates of Enzymatic Reactions

Consider the case of an enzyme-catalyzed reaction in which substrate S is converted to product P, catalyzed by enzyme E. We want to measure the rate of this reaction in the direction S <——> P. However, the reaction is reversible. If we begin with only S plus the enzyme, P will be formed. As the concentration of P increases and that of S decreases, the reverse reaction will take place, becoming more important as the concentration of P increases. Accordingly, the rate of the forward reaction must be measured quickly before any appreciable amount of P is formed. We call this quickly-measured forward reaction rate the initial velocity.

$$E + S \leftrightarrow ES \leftrightarrow EI \leftrightarrow EP \leftrightarrow E + P$$

Figure 2.2 Possible steps in the conversion of substrate S to product P under the influence of enzyme E. Three steps are identified: An initial enzyme-substrate complex (ES) is converted to an enzyme-intermediate complex (EI), and then the enzyme-product complex (EP) breaks down to product and free enzyme.

Effect of Substrate Concentration

Let us carry out an experiment in which we set up 10 test tubes, each containing a solution at 25°C, pH 7.0, and a fixed concentration of enzyme E. We add a specific amount of substrate S to each test tube, mix it, and quickly measure the rate of the reaction, either by measuring the rate of disappearance of S or the rate of appearance of P. Let us assume that the concentration of S in test tube one is two micromolar, with higher and higher concentrations in the other tubes such that the concentration of S in tube 10 is 500 micromolar. After getting our initial velocities, expressed in units of micromoles of S disappearing per minute (or micromoles of P appearing per minute), we plot the initial velocity as a function of substrate concentration. The graph should look like the one shown in Figure 2.3.

Figure 2.3 shows that initial velocity is higher as substrate concentration is increased. Note that the relationship is not linear but hyperbolic. That is, at low substrate concentration the velocity increases linearly with increasing substrate concentration, but at a higher substrate concentration the velocity flattens out, approaching a maximum velocity called V_{max}. If we determine the value of the maximum velocity, divide this in half, then determine what substrate concentration will produce one half of V_{max}, we get a concentration known as K_m. The K_m (also known as the Michaelis constant) is defined as the substrate concentration needed to produce one half the maximal velocity of an enzyme-catalyzed reaction. The K_m has units of concentration. V_{max} is the limiting rate for the velocity of an enzyme-catalyzed reaction at fixed enzyme concentration. It occurs when enzyme active sites are so totally saturated with substrate that as a substrate molecule is converted to product and leaves the enzyme, another substrate molecule immediately binds.

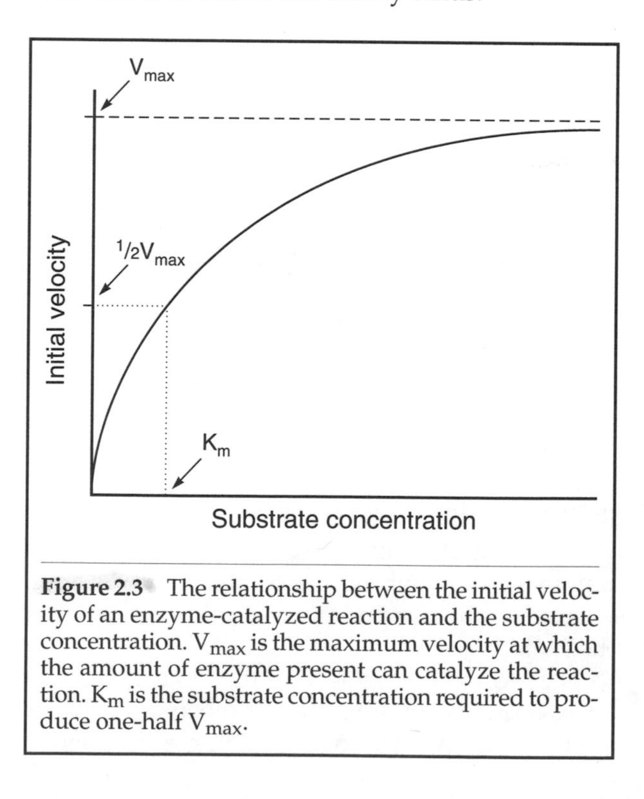

Figure 2.3 The relationship between the initial velocity of an enzyme-catalyzed reaction and the substrate concentration. V_{max} is the maximum velocity at which the amount of enzyme present can catalyze the reaction. K_m is the substrate concentration required to produce one-half V_{max}.

V_{max} and K_m are known as kinetic parameters for an enzyme-catalyzed reaction. K_m is said to reflect the affinity of an enzyme for its substrate; the smaller the value of K_m the greater the affinity an enzyme has for its substrate. The reverse reaction will have its own kinetic parameters, because the product becomes the substrate for the backward reaction. The kinetic parameters for the reverse reaction are unlikely to be identical to those for the forward reaction.

In the cell, the substrate concentration is generally equal to or less than the value for its K_m. This offers two advantages: (a) a substantial fraction of enzyme catalytic ability is being used; and (b) the substrate concentration is low enough that the enzyme can still respond to changes in substrate concentration because it is on the steep part of the curve (see Figure 2.3). If the substrate concentration is much greater than K_m we would get efficient use of the enzyme, but it would respond less effectively to changes in substrate levels because it is on the flat part of the curve (see Figure 2.3). On the other hand, if substrate concentration is normally much less than K_m, the enzyme is very sensitive to changes in substrate concentration, but the enzyme would not be used efficiently because there would be too much enzyme for the available substrate.

There are enzymes that catalyze the same reaction in different tissues with different kinetic parameters. Often products of different genes, they are known as isozymes or isoenzymes. When glucose enters a cell, the first thing that happens is that a phosphate group attaches to it. This reaction is catalyzed by four isozymes known as hexokinases I, II, III, and IV. Three of these isozymes have low K_m values for glucose (20-120 micromolar); the fourth (hexokinase IV, also known as glucokinase and found in the liver) has a high K_m for glucose (5 mM). The low K_m isozymes can phosphorylate glucose even when the blood concentration is low. This is especially important to the brain, which depends solely on glucose as its fuel under normal circumstances. The high K_m isozyme is found in the liver, where glucose is stored when blood concentration is elevated. The liver isozyme thus readily responds to glucose as a substrate only when blood glucose is elevated. Figure 2.4 plots the initial velocity and substrate concentration for a low K_m hexokinase and for glucokinase.

Determining reliable values for the kinetic parameters K_m and V_{max} from the velocity-substrate concentration curve shown in Figure 2.3 is difficult. A wide range of substrate concentrations must be tested to ensure a reasonable value of V_{max} because we also need this value to determine K_m. An easier way, shown in Figure 2.5, is a Lineweaver-Burk plot. Here, the reciprocals of velocity and substrate concentration are plotted, which allows one to determine accurately the kinetic parameters for an enzyme-catalyzed reaction because these values are determined from intercepts on the horizontal and vertical axes. One can thus obtain more accurate values for the kinetic parameters from fewer data points.

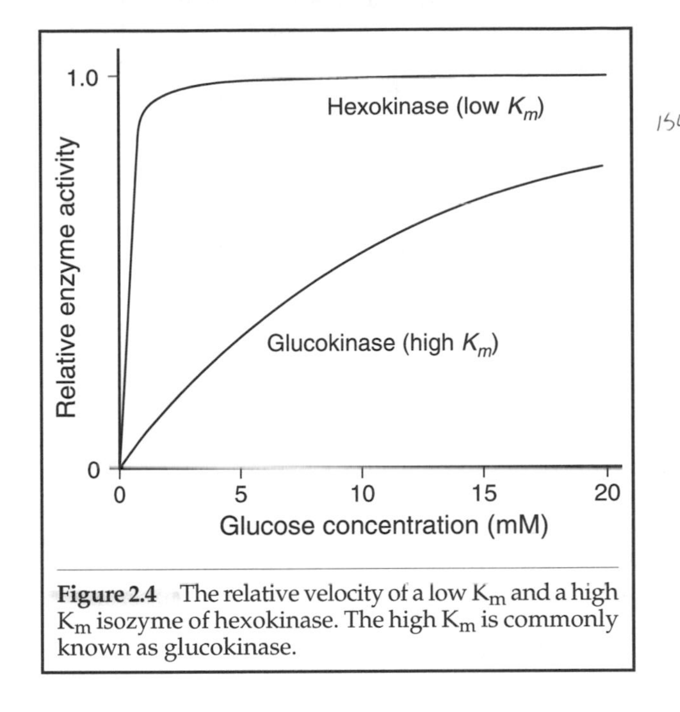

Figure 2.4 The relative velocity of a low K_m and a high K_m isozyme of hexokinase. The high K_m is commonly known as glucokinase.

isozymes

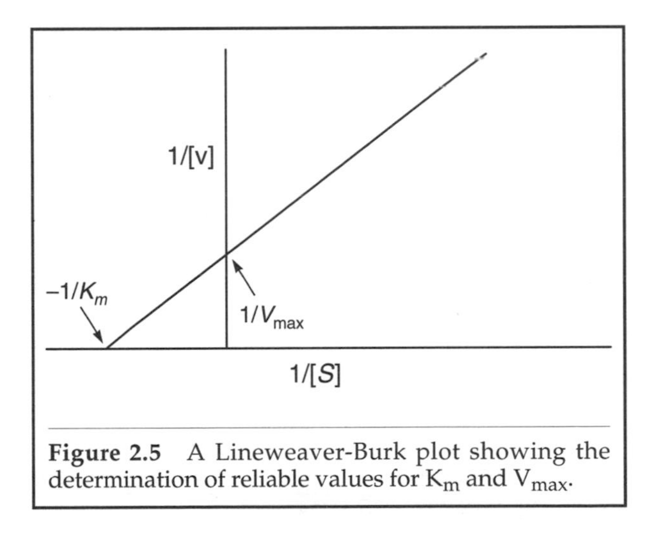

Figure 2.5 A Lineweaver-Burk plot showing the determination of reliable values for K_m and V_{max}.

Turnover Number (k_{cat})

Turnover number (k_{cat}) is the maximum number of molecules of substrate converted to product per enzyme active site per unit of time (usually one second). For enzymes, k_{cat} measures catalytic efficiency when they are saturated with substrate. Some enzymes are extremely efficient catalysts. For example, catalase, which breaks down hydrogen peroxide, has a k_{cat} of over one million. However, most enzymes do not operate under conditions in which they are constantly saturated with substrate. A better way to express their catalytic efficiency *in vivo* is to use the expression k_{cat}/K_m, which gives a truer picture under physiological conditions. To put this into perspective, the turnover number is like the maximum speed a car can travel. The ratio of k_{cat}/K_m is more like the everyday cruising efficiency of the car because one seldom drives at maximum speed.

Effect of Enzyme Concentration

If an enzyme is saturated with substrate, adding more enzyme increases the reaction velocity. V_{max} is thus proportional to enzyme concentration. However, changing enzyme concentration only increases V_{max} and has no influence on K_m. We can use the relationship between enzyme concentration and maximum velocity to determine how much of a particular enzyme is present in a tissue or fluid (e.g., blood).

Effect of Temperature on Enzyme Reactions

Like all chemical reactions, enzyme-catalyzed reactions increase in rate if the temperature is increased. However, since the forces holding the three-dimensional conformation of an enzyme are generally weak, heating too much disrupts the conformation and decreases its activity. Thus, if we plot the velocity of an enzyme-catalyzed reaction as a function of temperature, the curve rises with increasing temperature until about 50°C, at which point most enzymes start to denature, and the velocity drops quickly. Biochemists describe the relationship between temperature and reaction rate by the quotient Q_{10}, which describes the fold increase in reaction rate for a 10°C rise in temperature. For many biological processes, the rate of a reaction approximately doubles for each 10°C increase in temperature. This means the Q_{10} is about two.

Effect of pH

Most enzymes have a pH optimum, that is, a particular pH or narrow pH range where enzyme activity is maximal. Values of pH on either side of optimum produce lower reaction rates. The reason for this is that some of the forces holding an enzyme in its native conformation depend on charged groups. A change in pH can alter these charges, thus attenuating enzyme function. The change may directly influence the active site, or some other part of the enzyme that indirectly affects the active site. A change in pH may also alter the substrate for an enzyme, which could also influence rate. For most enzymes, their pH optimum reflects where they are active in the body. For example, the stomach enzyme pepsin has a pH optimum around 2 because the stomach is acidic.

Enzyme Inhibition

Enzymes can be inhibited by a variety of substances. We describe these inhibitors according to how they influence the enzyme. Competitive inhibitors resemble the normal substrate for an enzyme, for they bind to the active site but cannot be changed by the enzyme. The inhibitor simply occupies the active site, binding, leaving, binding, etc., in a reversible fashion. Competitive inhibitors thus compete with the normal substrate for a place on the active site of the enzyme. This inhibition can be overcome by adding excess substrate. Accordingly, a competitive inhibitor will not affect the V_{max} of the enzyme but will increase the K_m. The Lineweaver-Burk plot in Figure 2.6 shows the effect of a competitive inhibitor.

A noncompetitive inhibitor does not resemble the normal enzyme substrate and does not bind to the active site. However, when bound to the enzyme, it interferes with its function. Hence, noncompetitive inhibitors lower V_{max} but do not alter K_m. Examples of competitive inhibitors are heavy metal ions (e.g., Hg^{2+}, Pb^{2+}). Figure 2.6 also shows how a noncompetitive inhibitor influences the Lineweaver-Burk plot. Many drugs are based on the principles of enzyme inhibition, where specific reactions in bacteria, viruses, or tumors are targeted by the drug.

Provision of Reactive Groups by Cofactors

Enzymes have reactive groups in the form of side chains of amino acids as well as the N and C termini. However, they may need other reactive groups not available on amino acids, called cofactors, in order to carry out their functions as catalysts. Cofactors may be metal ions, such as Zn^{2+}, Mg^{2+}, or Mn^{2+} ions, or they may be organic molecules called coenzymes. The polypeptide part of the enzyme is called the apoenzyme; when combined with the cofactor, we have the holoenzyme, illustrated as follows:

$$\text{apoenzyme} + \text{cofactor} = \text{holoenzyme}$$

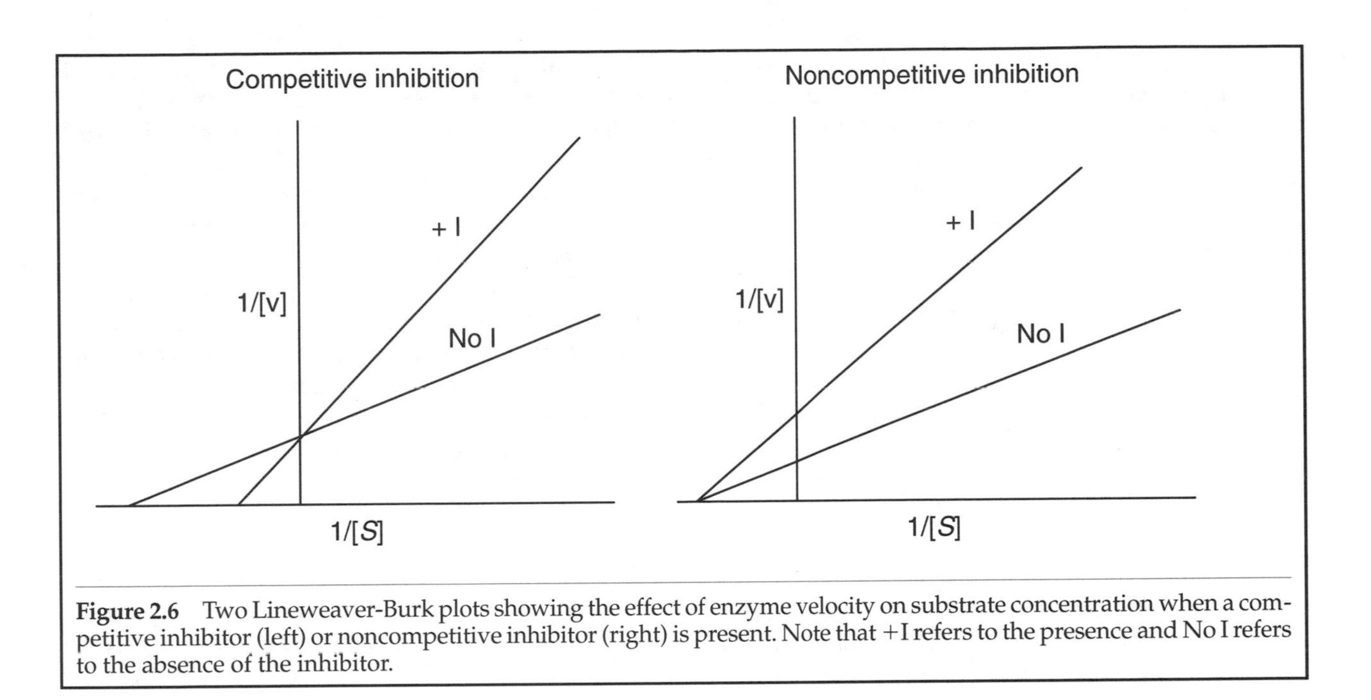

Figure 2.6 Two Lineweaver-Burk plots showing the effect of enzyme velocity on substrate concentration when a competitive inhibitor (left) or noncompetitive inhibitor (right) is present. Note that +I refers to the presence and No I refers to the absence of the inhibitor.

A cofactor tightly bound to the enzyme at all times is called a prosthetic group. If the cofactor is not tightly bound but combines with the enzyme and the other substrate during the reaction, we consider it a second substrate.

There are eight B vitamins which we need because they form the basic components for coenzymes. Humans and many animals eating a plant and animal diet have lost the ability to synthesize the B vitamins, so they must come from the diet. Table 2.1 lists the eight B vitamins, the coenzymes they form, and their short form names.

People often stress the importance of vitamins but neglect the minerals. We need many more minerals in our diet than vitamins. Many minerals affect the function of enzymes. For example, zinc is an essential component of more than 120 different enzymes, including those for synthesizing RNA and DNA, pancreatic digestive enzymes, and enzymes involved in carbohydrate, fat, protein, and alcohol metabolism.

Oxidations and Reductions

Oxidation-reduction, or redox, reactions are extremely important for all organisms. In these reactions, something gets oxidized, and something gets reduced. You may have encountered in earlier chemistry courses the term oxidation, which means something loses electrons, and reduction, which means something gains electrons. The familiar memorization tool that Leo (oxidation loses electrons) says Ger

Table 2.1 The B Vitamins, the Coenzymes They Form, and Their Common Abbreviations

B vitamin	Coenzyme	Abbreviation
Thiamin (B_1)	Thiamin pyrophosphate	TPP
Riboflavin (B_2)	Flavin adenine dinucleotide	FAD
Niacin	Nicotinamide adenine dinucleotide	NAD
Vitamin B_6	Pyridoxal phosphate	PLP
Pantothenic acid	Coenzyme A	CoA
Folate (folacin)	Tetrahydrofolic acid	THFA
Biotin	Biotin	–
B_{12}	Cobalamin coenzymes	–

(reduction gains electrons) has been used by students for many years. Redox reactions are absolutely critical to life, and underlie all aspects of metabolism. In all cases the electron gain (reduction) is directly connected with electron loss (oxidation), as seen in the reaction

$$Fe^{3+} + Cu^{+} \longrightarrow Fe^{2+} + Cu^{2+}$$

In this example, the iron (Fe^{3+}) is reduced because it gains an electron from the copper to form (Fe^{2+}), whereas the copper (Cu^{+}) is oxidized by losing an electron to the iron, thereby becoming Cu^{2+}.

In the cell, there are redox reactions where ions are oxidized and reduced as the above equation shows. However, in many important redox reactions it is not easy to see that electrons are lost from one molecule and gained by another. For example, Figure 2.7 shows two dehydrogenations in which two hydrogens are lost from the two partial structures. The important thing is that when the hydrogens leave, they exit with electrons. In the first example, two hydrogen atoms are lost, and a double bond is generated. In the second reaction, two hydrogens are lost from a secondary alcohol, generating a ketone group. In this latter case, one hydrogen comes off as a hydride ion and the other as a proton. The hydride ion is simply a hydrogen atom (with one proton and one electron) with an additional electron so that it has a negative charge. In summary, dehydrogenations are oxidation reactions in which electrons are lost as part of hydrogen atoms or hydride ions.

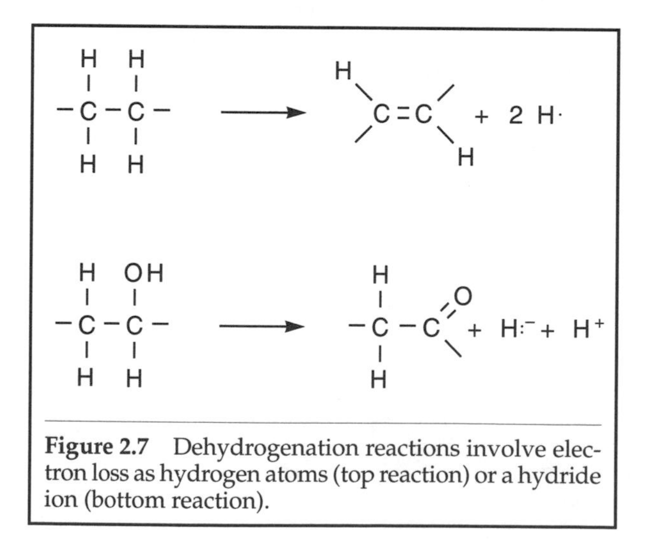

Figure 2.7 Dehydrogenation reactions involve electron loss as hydrogen atoms (top reaction) or a hydride ion (bottom reaction).

Figure 2.7 shows dehydrogenation, but these reactions are also accompanied by hydrogenation reactions, which means a molecule is reduced because it accepts two hydrogen atoms (each with an electron) or a hydride ion (with two electrons).

Two major coenzymes are involved in most cell redox reactions. The coenzyme flavin adenine dinucleotide (FAD; see Table 2.1) can accept two hydrogen atoms to become $FADH_2$ (the reduced form of FAD). FAD is the oxidized form of the coenzyme and $FADH_2$ the reduced form. The FAD coenzyme is involved in redox reactions where two hydrogen atoms are removed from chemical structures, such as the top reaction in Figure 2.7. Those shown in the lower part of Figure 2.7 involve the NAD^+ coenzyme, which accepts a hydride ion, becoming NADH.

Figure 2.8 shows two examples of redox reactions that use the FAD and NAD^+ coenzymes. In the first reaction, part of the citric acid (Krebs) cycle, succinate is oxidized to fumarate by losing two hydrogen atoms, forming the double carbon-to-carbon bond. The two hydrogen atoms, each carrying an electron, are picked up by the coenzyme FAD, forming $FADH_2$. Thus succinate is oxidized to fumarate, and FAD is simultaneously reduced to $FADH_2$ in a redox reaction. In the lower example, lactate loses a hydride ion and a proton when oxidized to pyruvate. At the same time, NAD^+ accepts the hydride ion, becoming the reduced form of the coenzyme NADH. Most redox reactions are reversible, so pyruvate can be reduced to lactate or fumarate reduced to succinate.

The net direction of redox reactions depends on the relative concentrations of the oxidized and reduced forms of the substrates and coenzymes. During exercise, when the rate of pyruvate formation increases in muscle, the enzyme lactate dehydrogenase produces lactate, attempting to maintain equilibrium. The lactate can travel from the muscle cell to the blood, where its concentration is often used as an indicator of the exercise intensity. Notice in Figure 2.8 that we write L-lactate because the middle carbon is chiral, and the L refers to the absolute configuration of the lactate. Lactate comes from lactic acid, a modestly strong acid (pK_a about 3.1). Thus, at the neutral pH where this reaction occurs, lactic acid exists as the ion lactate because it has lost its proton. The same can be said for pyruvate, from pyruvic acid.

Isozymes (Isoenzymes)

Isozymes, or isoenzymes, are different molecular forms of the same enzyme, catalyzing the same reaction with the same mechanism but with different kinetic parameters. Isozymes are typically oligomeric enzymes, or proteins with quaternary structure (i.e., containing subunits). Different subunits yield different kinetic parameters. A typical example of isozymes is the enzyme lactate

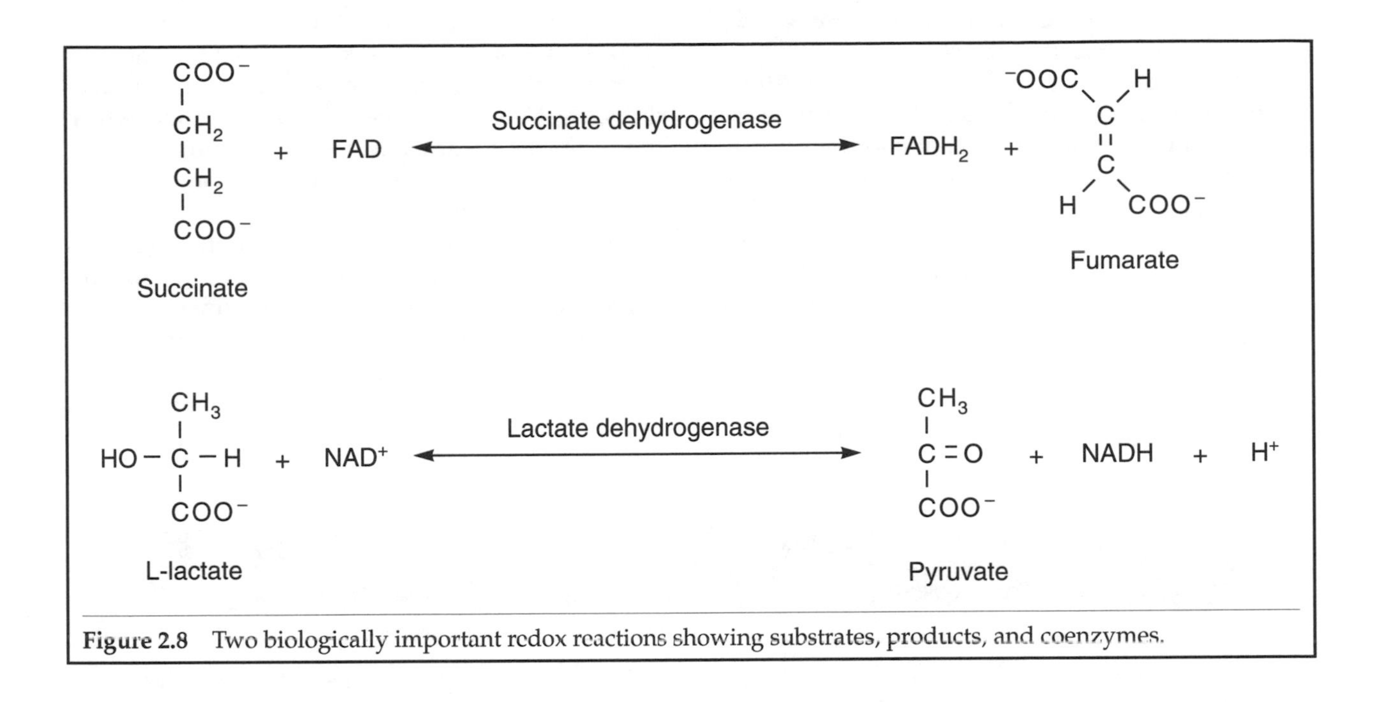

Figure 2.8 Two biologically important redox reactions showing substrates, products, and coenzymes.

dehydrogenase (LDH). LDH is a tetramer, consisting of four subunits of two different kinds: H (standing for heart) and M (standing for muscle). Each is the product of a separate gene. The five different kinds of LDH isozymes are

H_4	H_3M	H_2M_2	HM_3	M_4
LDH−1	LDH−2	LDH−3	LDH−4	LDH−5.

Isozymes 1 and 2 predominate in heart muscle and slow twitch skeletal muscle fibers and favor oxidation of lactate, whereas isozymes 4 and 5 favor reduction of pyruvate and are found mainly in fast twitch skeletal muscle fibers.

Another enzyme that exists in the form of isozymes is creatine kinase (we will discuss what it does later). It has two different kinds of subunits: M for muscle and B for brain. Creatine kinase exists as a dimer. Thus we have the isozymes MM (mainly in skeletal muscle), MB (mainly in heart), and BB (mainly in brain). Measurement of the concentration of specific creatine kinase isozymes in blood reveals whether skeletal muscle or heart muscle damage has occurred. For example, a heart attack damages the heart muscle and increases the concentration of the MB isozyme in the blood.

The contractile protein myosin is found in a variety of tissues but mainly in skeletal, heart, and smooth muscle. Each myosin molecule is a hexamer, containing two heavy chains and four light chains (see Figure 1.15). The many different forms of myosin depend on the tissue or cell type where it is found.

Measuring Enzyme Activity

We sometimes need to know how many functional enzyme molecules exist in a fluid (e.g., blood) or tissue. Because the number of molecules of functional enzyme is proportional to the V_{max}, measurement results are in units of reaction velocity per unit weight of tissue, per unit amount of protein, or per volume of fluid. Examples are micromoles of product per gram of tissue per minute, millimoles of substrate disappearing per milligram of protein per minute, or micromoles of product formed per milliliter of blood per minute, respectively. The expression of the V_{max} or, as it is commonly called, the activity of the enzyme is important in a variety of physiological and clinical conditions. For example, the activity of mitochondrial enzymes can almost double given the appropriate exercise training stimulus. We may want to determine if a particular exercise training program alters the metabolism of a muscle by measuring the activity of selected enzymes. We can also tell if a particular tissue is damaged by measuring the activity of isozymes specific for that tissue that are released due to cell membrane damage.

When measuring the activity of an enzyme certain principles must be established and rigidly followed. First, we need to make the measurements at a substrate (or substrates) concentration high enough to generate a true V_{max}. The pH of the reaction and the temperature at which it is measured should also be standardized so that meaningful comparisons can be made. Finally, we need to have a simple method for measuring the disappearance of substrate or appearance of product.

This determination can be done if the substrate or product is colored or can be made to generate a colored complex. For example, phosphate appearance can be readily measured because it forms a colored complex with a number of agents. One useful technique takes advantage of two properties of the coenzyme NADH. First, NADH absorbs light at 340 nm wavelength, so we can measure its rate of formation or disappearance with a spectrophotometer. The relationship is as follows: A 0.1 mM solution of NADH has an absorbance of 0.627. Second, NADH fluoresces when bombarded with light of a specific wavelength; thus we can measure its appearance or disappearance with a fluorometer. If the specific reaction does not actually involve NADH, it can be connected to a reaction which does. The rate of the reaction in question then dictates the rate of a connection reaction, in which NADH is formed or lost.

Biochemists use the term International unit (designated as U) to express the activity of an enzyme. One U of enzyme activity is the amount of enzyme that converts one micromole of substrate to product in one minute. Thus if an enzyme has an activity of 15 U per gram, 15 micromoles of product form per minute per gram of tissue.

Summary

Enzymes are biological catalysts—specialized proteins that speed up reactions in cells. Highly specific, they catalyze reactions involving single substrates or a closely related group of substrates. Enzymes have a Michaelis constant, K_m, which is the substrate concentration needed to produce one half the maximal velocity (V_{max}) of the enzyme reaction. The K_m, a characteristic constant for an enzyme, reflects inversely the affinity of the enzyme for its substrate. The maximal velocity of an enzyme-catalyzed reaction, or V_{max}, is proportional to the amount of enzyme present and can only be determined when the enzyme is saturated with its substrate. Measurement of the V_{max} thus determines the amount of enzyme present. An International unit (U) is defined as the amount of enzyme needed to convert one micromole of substrate to product in one minute. The action of enzymes can be hindered by the presence of inhibitors—specific substances that resemble the normal substrate and compete with it (competitive inhibitors) or that irreversibly alter the structure of the enzyme (noncompetitive inhibitors).

Many enzymes require the presence of nonprotein substances to function. These cofactors can be organic molecules, that is coenzymes, or they can be metal ions. Most coenzymes are derived from the B vitamins in our diet, while our need for many specific mineral nutrients relates to their role as enzyme cofactors. Isoenzymes (isozymes) are closely related, but different enzyme molecules that catalyze the same reaction but differ in certain properties, such as K_m. Some of the most important enzymes are the dehydrogenases, which add or remove electrons from their substrates. They play a major role in producing energy in the process of oxidative phosphorylation.

Chapter 3

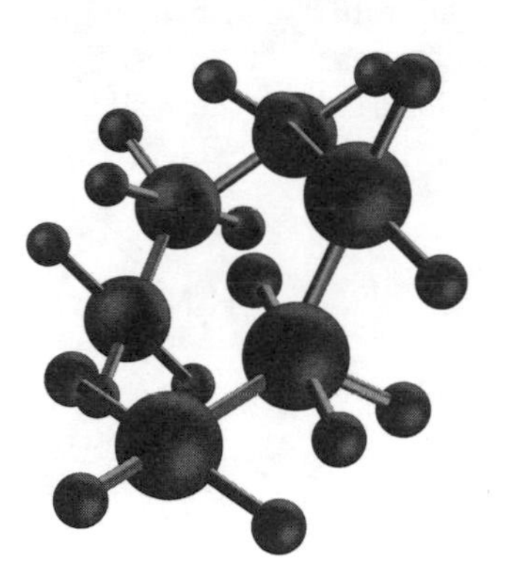

DNA and RNA

DNA (deoxyribonucleic acid) and RNA (ribonucleic acid) are polynucleotides—large molecules (polymers) composed of individual nucleotides (monomers) joined end to end. The monomers themselves are made up of three main components: bases, sugars, and phosphoric acid.

The Components

Five main bases are found in polynucleotides that are further subdivided into the purines and the pyrimidines. The two purines are adenine, represented by A, and guanine (G). The pyrimidines are cytosine (C), thymine (T), and uracil (U). Thymine is only found in DNA and uracil only in RNA. Figure 3.1 shows the chemical structures for the pyrimidines and purines.

Two different sugars exist in polynucleotides; deoxyribose, found only in DNA, and ribose, found in RNA. Figure 3.1 also shows these chemical structures. Note that the numbering system for these sugars has a prime (′) to differentiate the sugar numbers from the numbers associated with the carbon and nitrogen atoms in the ring portion of the bases (not shown). Phosphoric acid is found in nucleotides and polynucleotides, but because the pH of the cell is near 7, it is ionized and exists as phosphate (the ionized form of phosphoric acid).

Nucleosides are formed when a base joins to a sugar (see Figure 3.2). Thus we have deoxynucleosides (if the sugar is deoxyribose) or simply nucleosides (if the sugar is just ribose). When a specific base joins to a sugar, the name of the resulting nucleoside (or deoxynucleoside) reflects the name of the base. For example, adenine plus ribose or deoxyribose gives adenosine or deoxyadenosine. Thymine plus deoxyribose gives deoxythymidine—sometimes called thymidine because thymine only joins to deoxyribose. The other nucelosides are guanosine, cytidine, and uridine.

Nucleotides are formed when a nucleoside is joined to a phosphoric acid (actually a negatively charged phosphate group). The phosphate is normally attached to one of the hydroxy (OH) groups on the sugar. An acid attached to an OH (or alcoholic group) is an ester; if the acid is derived from phosphoric acid (i.e., phosphate), the ester is a phosphate ester. The location of the phosphate is designated by the number of the sugar carbon it is attached to via the ester bond. Figure 3.2 shows an example of a nucleotide. It is very common to have two or three phosphate groups attached together and then to the sugar. Thus we can have nucleotides with one phosphate (i.e., a nucleoside monophosphate—NMP), with two phosphates (i.e., a nucleoside diphosphate—NDP), or with three phosphates (i.e., a nucleoside triphosphate—NTP). A nucleotide

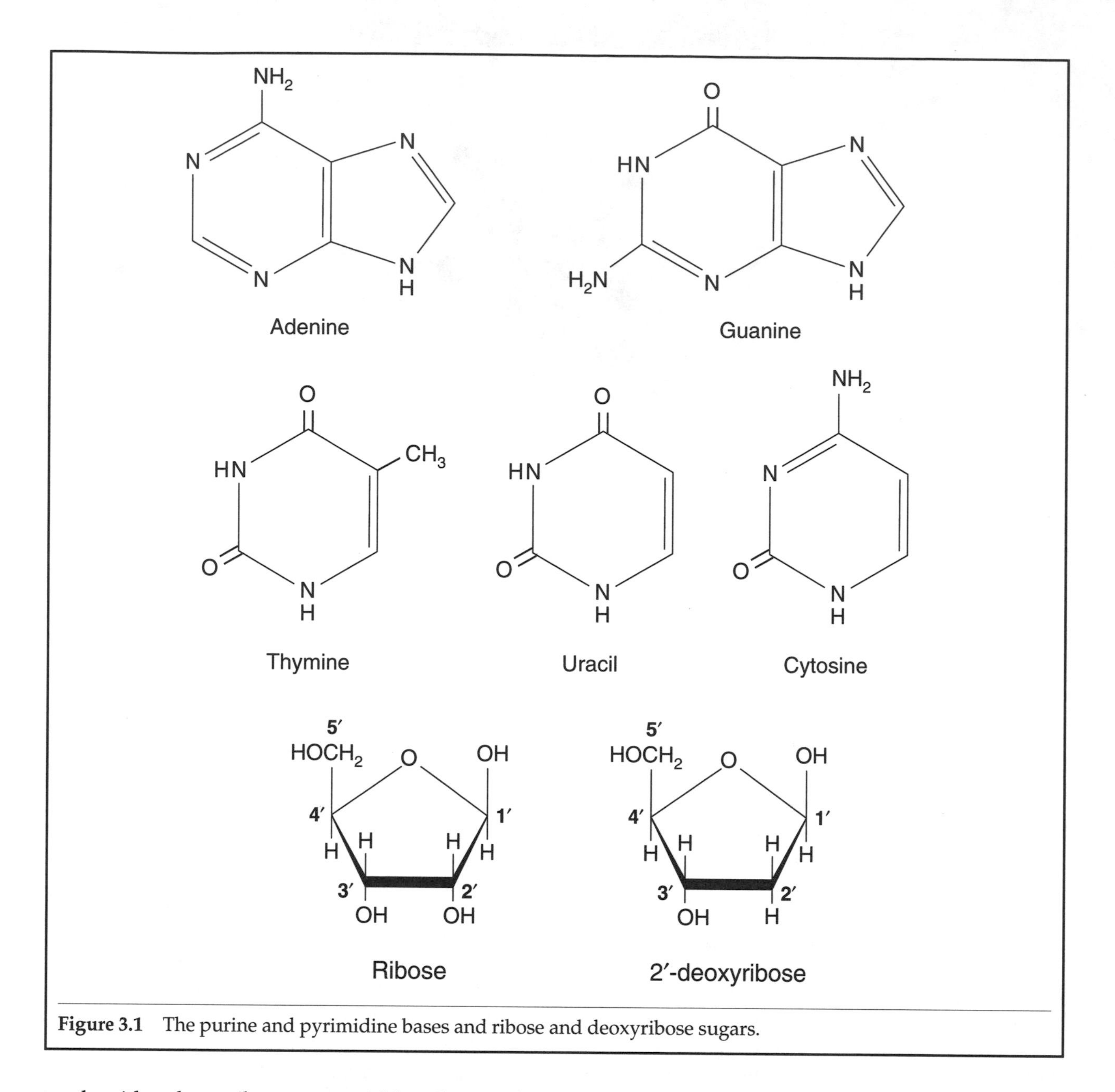

Figure 3.1 The purine and pyrimidine bases and ribose and deoxyribose sugars.

made with a deoxyribose sugar yields a deoxynucleoside mono- (dNMP), di- (dNDP) or tri- (dNTP) phosphate.

So far, we have described nucleotides in general terms. If we know the base that is used, we can be more specific. Suppose we have a nucleotide composed of the base adenine, the sugar ribose, and a phosphate group attached to the 5′ (the five prime) carbon of ribose. It would be adenosine 5′-monophosphate, abbreviated as 5′-AMP (sometimes simply AMP). If two phosphates are attached together and then attached to the 5′ position on ribose, it would be adenosine 5′-diphosphate or 5′-ADP (or simply ADP). Three phosphate groups attached to the 5′ position of guanosine (the nucleoside) would be known as guanosine 5′-triphosphate or 5′-GTP (or simply GTP).

If two phosphates are joined together, as in 5′-ADP or 5′-ATP, we have an anhydride bond—a bond joining two acid groups together. Such bonds are energy-rich because their hydrolysis (i.e., splitting) releases much energy. We will discuss hydrolysis later.

DNA

When a human egg cell, containing 23 chromosomes, is fertilized by a sperm cell, containing 23 chromosomes, the resulting fertilized egg, containing 46 chromosomes, undergoes repeated

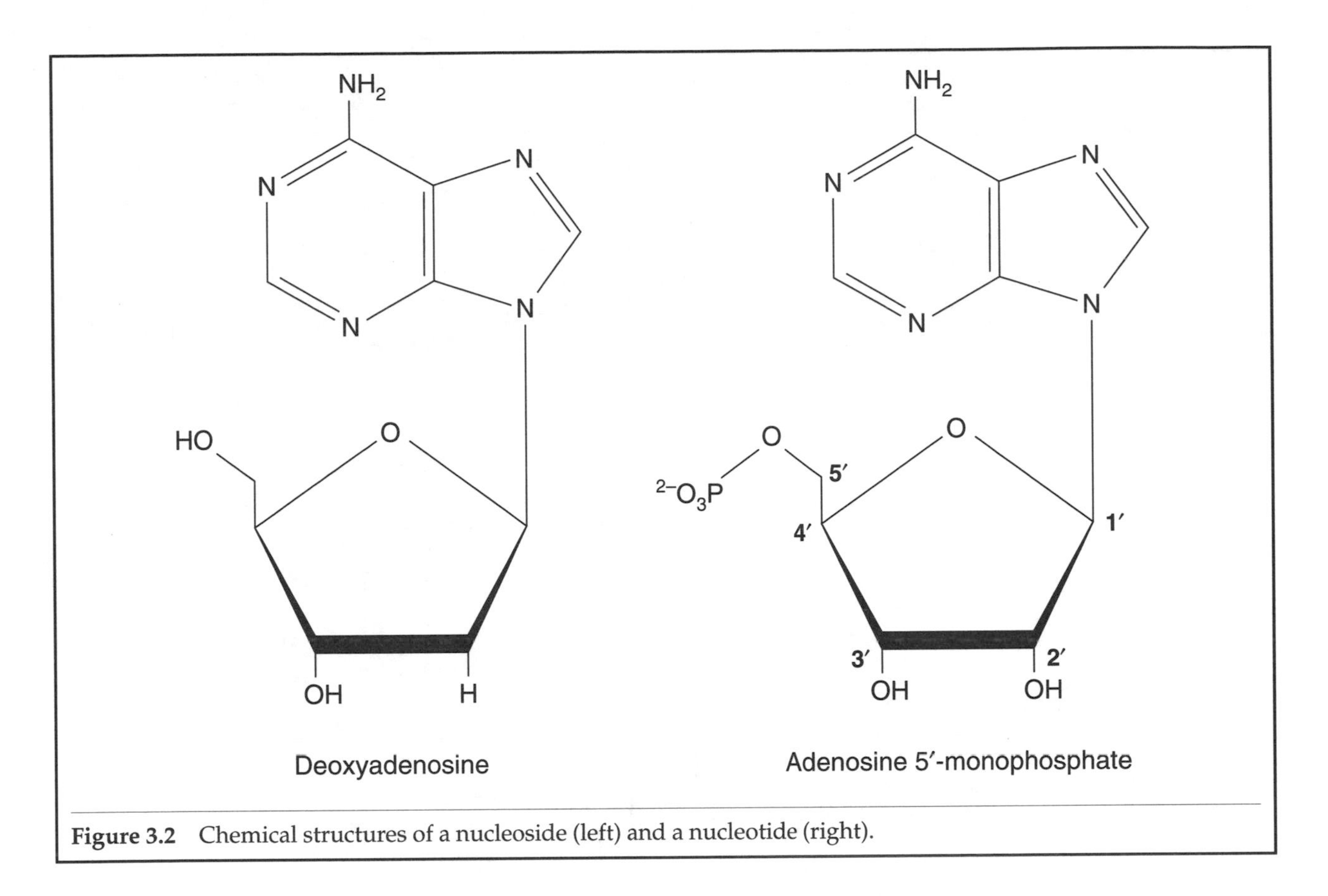

Figure 3.2 Chemical structures of a nucleoside (left) and a nucleotide (right).

cycles of cell division (mitosis). Cells undergo differentiation into specialized cells, which further increase in number through cell division, eventually generating a human baby. By adulthood, there are about 10^{13} nucleated cells of many different kinds, formed into specialized tissues and about 2×10^{13} non-nucleated red blood cells. The nucleus in each cell (some cells, e.g., muscle cells, contain more than one nucleus) contains 46 chromosomes: two copies of chromosomes 1 to 22 plus two X chromosomes (female) and two copies of chromosomes 1 to 22 plus one X and one Y (male). Egg and sperm cells contain only 23 chromosomes: Egg cells contain a single copy of chromosomes 1 to 22 plus an X; sperm cells contain a single copy of chromosomes 1 to 22 plus either an X or Y. Thus the sperm that fertilizes the egg determines the sex of the offspring.

Each chromosome contains one very large DNA molecule, so in a somatic cell nucleus (non-sex cell) there are 46 DNA molecules but only 23 DNA molecules in a sperm or egg nucleus. Figure 3.3 shows part of the structure of a DNA molecule. On the left is part of a single DNA strand indicating the joining of the deoxyribose sugars via phosphate groups. Note that the sugars are linked together by phosphate groups attached to the 3′ carbon of one sugar and then to the 5′ carbon of the next. The base is attached to the 1′ carbon of each deoxyribose. On the right side of Figure 3.3 is double-stranded DNA with the two strands (or chains) held together by hydrogen bonding between the bases. In all cases, an adenine base (A) in one chain is linked to a thymine (T) base in the other with two hydrogen bonds, and a cytosine (C) base in one strand is joined to a guanine (G) base in the other with three hydrogen bonds. The two chains are thus complementary, if we know the base sequence in one chain, we know the base sequence in the other because A hydrogen bonds to T and C hydrogen bonds to G. The two chains are also antiparallel. This means that one chain runs 5′ to 3′ from top to bottom, whereas the other runs 3′ to 5′ from top to bottom. Finally, the two chains are coiled around a common axis forming the familiar double helix, with a full turn of the helix every 3.4 nm (see Figure 3.4). Each full turn of the helix has 10 stacked pairs of bases.

DNA molecules are very large and described according to the number of base pairs. Thus terms like base pairs (bp), thousands of base pairs (kilobase pairs—kb), or megabase pairs (Mb) are widely used. In summary, DNA is double stranded, held together by hydrogen bonds between bases, complementary (A = T and G = C), antiparallel—the two chains run in opposite directions, and twisted into a double helix.

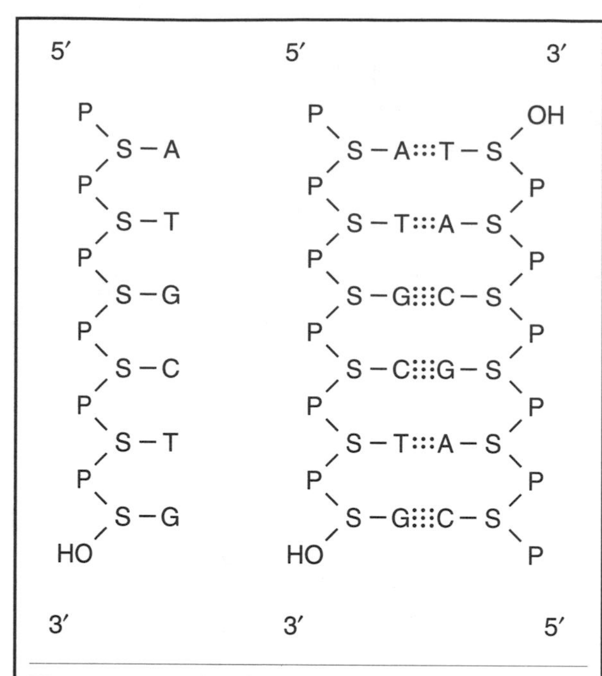

Figure 3.3 A shorthand way of illustrating single-strand DNA (left) and double-stranded DNA (right). S refers to deoxyribose sugar; P represents the phosphate groups. The sugar phosphate backbone runs from the 5′ to the 3′ end. The complementary strand of DNA on the right is antiparallel. The double-stranded DNA is joined by hydrogen bonds (dotted lines), two join the bases adenine (A) and thymine (T) and three join the bases cytosine (C) and guanine (G).

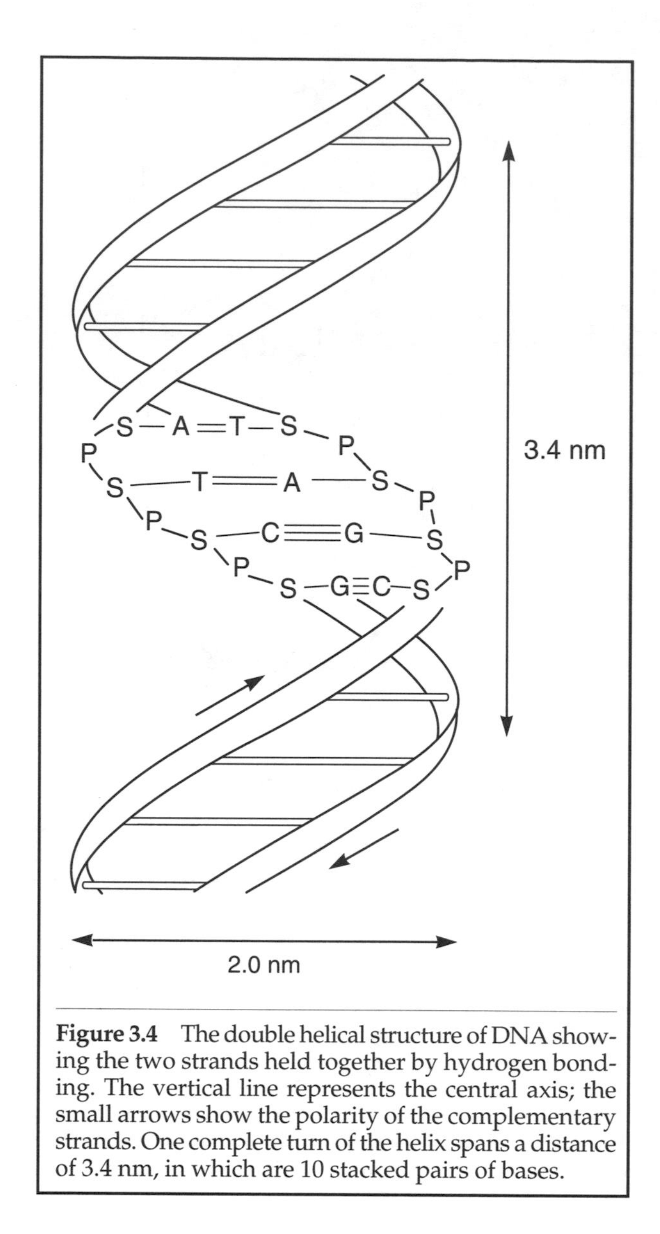

Figure 3.4 The double helical structure of DNA showing the two strands held together by hydrogen bonding. The vertical line represents the central axis; the small arrows show the polarity of the complementary strands. One complete turn of the helix spans a distance of 3.4 nm, in which are 10 stacked pairs of bases.

DNA Replication

Before a parent cell divides to produce offspring cells, known as daughter cells, the DNA must first be duplicated. Each daughter cell gets an exact copy of the parent DNA. To duplicate the parent DNA, each strand of the parent DNA acts as a template on which a complementary copy is made. Each daughter cell then receives one strand of the parent DNA and a newly synthesized complementary copy. This form of DNA replication is called semi-conservative.

DNA replication is catalyzed by DNA polymerase. The precursors needed to make a complementary copy of DNA are deoxyribonucleotide triphosphates (dNTP): dATP, dCTP, dGTP, and dTTP. Since DNA is so large, replication takes place simultaneously at multiple sites on the same molecule. DNA polymerase unwinds the strands where replication is to begin, and a complementary copy is made of each strand by adding nucleotides one at a time, using the dNTP precursors. The incoming dNTP attaches to the 3′OH group of the preceding nucleotide via a phosphate ester bond, and a pyrophosphate molecule (PPi) is released. The next dNTP comes in, attaches to the 3′OH of the preceding nucleotide and another PPi is released. In all cases, the correct dNTP comes in because of complementary base pairing. Figure 3.5 shows this schematically. Although both are duplicated, this figure shows only one of the two parent strands being copied.

As you can see, DNA polymerase reads the template strand in the 3′ to 5′ direction, whereas the newly synthesized strand grows in the 5′ to 3′ direction. Magnesium ions (Mg^{2+}) are needed to help the DNA polymerase enzyme function properly.

RNA

RNA or ribonucleic acid is a polynucleotide that differs from DNA in the following characteristics:

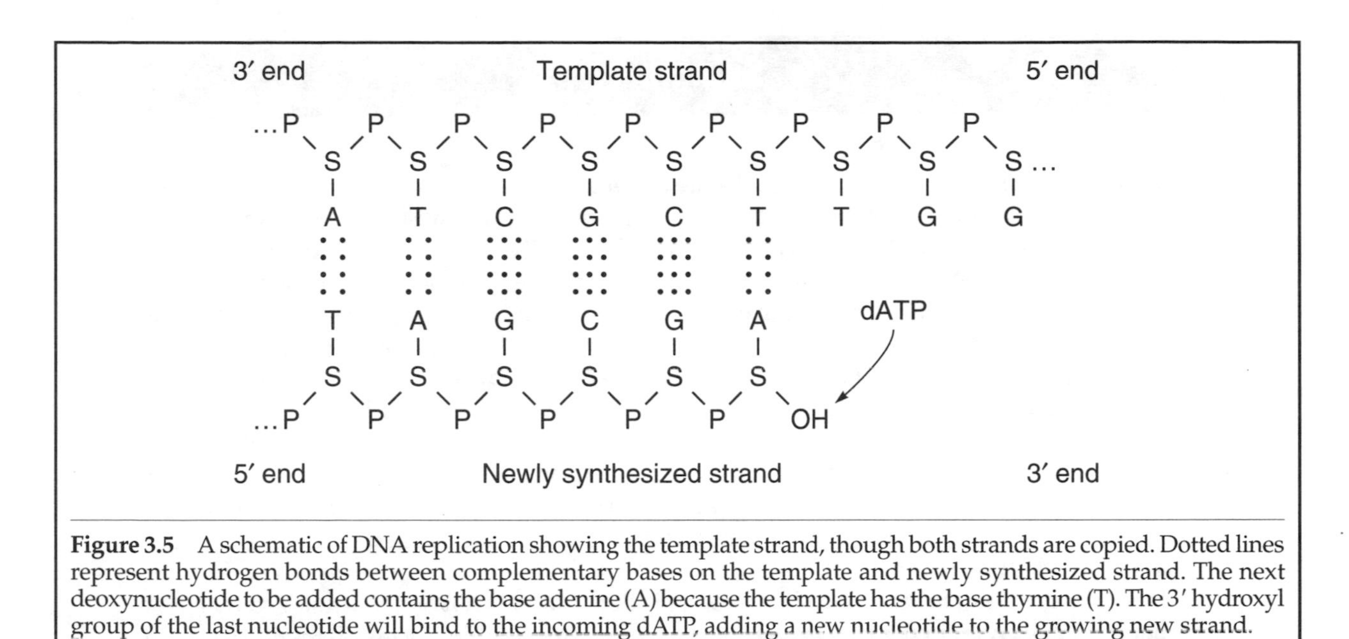

Figure 3.5 A schematic of DNA replication showing the template strand, though both strands are copied. Dotted lines represent hydrogen bonds between complementary bases on the template and newly synthesized strand. The next deoxynucleotide to be added contains the base adenine (A) because the template has the base thymine (T). The 3′ hydroxyl group of the last nucleotide will bind to the incoming dATP, adding a new nucleotide to the growing new strand.

(a) It is single-stranded; (b) it contains the sugar ribose instead of deoxyribose; (c) instead of the base thymine, it contains the base uracil (U); and (d) RNA molecules are much smaller than DNA molecules. However, RNA molecules can still be fairly large, and we often discuss their size by reference to the number of bases. For example, RNA can be 640 or 2,500 bases long; the latter would be described as 2.5 kb. RNA size is also commonly described in terms of the number of nucleotides (Nt); that is, saying 2,000 bases is equivalent to saying 2,000 Nt.

DNA constitutes the master file of genetic information, unfailingly reproduced over successive generations of cell division. RNA molecules are working copies of parts of DNA and thus tools used for making proteins. DNA replication occurs with high fidelity, or very few errors, for error in base sequence can be very harmful. Nonetheless, changes or mutations can and do occur. Mutations may be silent—that is, not expressed—or visible—that is, expressed. Visible mutations can be harmful. Mutations may be classified as base substitutions, in which a base is altered or substituted for by a closely related base, or frame shifts, in which a base is added or deleted. Either totally alters the DNA message in that region, if expressed in the cell where it occurs. Mutations occur through the action of harmful chemicals, known as mutagens, or from ultraviolet (UV) radiation, x-rays, or gamma rays.

We have seen that single strands of DNA can be joined together by hydrogen bonds if the bases in the two strands are complementary. We call this process hybridization, and it is also possible to get RNA-DNA hybridization and RNA-RNA hybridization. This property of hybridization of single-stranded polynucleotide chains is extremely important. For example, one can seek out a particular base sequence in single-stranded DNA or RNA if one has a short length of radioactively-labeled polynucleotide of complementary base sequence, that is, a probe. Hybridization utilizing specific probes underlies the techniques of Southern blotting (searching for specific DNA base sequences) and Northern blotting (seeking RNA base sequences).

Some viruses, known as retroviruses, contain single-stranded RNA instead of double-stranded DNA as their genetic material. When these viruses enter a cell, they bring with them an enzyme known as reverse transcriptase. Inside the host cell, reverse transcriptase catalyzes the formation of a strand of DNA complementary to the RNA molecule of the virus. Accordingly, reverse transcriptase is also known as RNA-directed DNA polymerase. The resulting RNA-DNA hybrid is then acted upon by an enzyme, known as a ribonuclease (or RNA degrading enzyme), that breaks down the RNA strand. The action of ribonuclease leaves a single strand of DNA, complementary to the virus RNA. DNA polymerase is then used to make a complementary copy of the single DNA strand, producing double-stranded DNA, which is then incorporated into the chromosomal DNA of the host. The AIDS virus is an example of a retrovirus. Reverse transcriptase is also an important laboratory enzyme, used to make DNA copies of RNA molecules, known as complementary or cDNA.

Summary

Deoxyribonucleic acid, or DNA, is the molecule of heredity. Each human cell nucleus contains 46 DNA molecules which are associated with protein, making up our 46 chromosomes. DNA consists of two polynucleotide chains. Each nucleotide in a DNA polynucleotide chain consists of an organic base attached to a sugar known as deoxyribose, containing a covalently bound phosphate group. Adjacent nucleotides in each chain are held together by phosphate ester bonds joining sugar molecules and phosphate groups. The bases in DNA are adenine (A), guanine (G), cytosine (C), and thymine (T). The two polynucleotide chains in DNA are held together by hydrogen bonds between complementary bases, A bonding with T and G bonding with C. The two chains are antiparallel, with one running in the 5′ to 3′ direction and the other 3′ to 5′; these directions describe the orientation of the sugar phosphate backbone of the polynucleotide. The complementary polynucleotide chains are wound in a double helix. Information for the sequence of amino acids in polypeptide chains is provided by the sequence of the four bases in each polynucleotide chain.

Before cells divide to form new cells, the DNA is duplicated in a reaction catalyzed by DNA polymerase. Each chain in the parent cell acts as a template on which a complementary strand of new DNA is made. Subsequently, offspring cells each receive a strand of parent DNA and a new complementary copy, such that daughter DNA and parent DNA are identical. Ribonucleic acid or RNA is a single-stranded polynucleotide, a copy of part of a DNA molecule known as a gene. RNA differs from DNA in that the sugar ribose replaces deoxyribose and the base uracil (U) replaces thymine. RNA information is used to make proteins, a topic covered inPart II.

Basic Molecular Biology

The phenotype of an organism represents its physical, observable characteristics, whereas the genotype represents the genetic factors responsible for creating the phenotype. The genome refers to the complete set of hereditary information in a species, or all the DNA in a nucleus. The genome is the same in every cell nucleus in an organism with the exception of the sex cells, which have half the number of chromosomes. Although it represents all the genes in all the chromosomes in the cell nucleus, not all the genome may be expressed.

The word gene is familiar to most people, yet it can have a variety of definitions. From a classic biology perspective it may be described as the basic unit of inheritance. From a modern perspective, it may be described as the segment of DNA that provides the information for the amino acid sequence of a polypeptide chain, because most genes provide this information. Lewin (1990) defines a gene as "the segment of DNA involved in the production of a polypeptide chain..." In humans, less than 5% of all DNA is actually expressed, or about 100,000 genes. Put another way, of the 3 billion base pairs making up the 23 (for women) or 24 (for men) different DNA molecules in the nucleus, most are never used to make protein molecules.

In this section, we will trace the path of gene expression from DNA base sequence to the making of a protein. In chapter 4, we will learn about the different kinds of RNA, the genetic code, gene transcription, and how transcription is regulated. In chapter 5, we will focus on the modification of the primary gene product to produce messenger RNA and how the information in messenger RNA is used to make a protein in the process of translation. We will pay particular attention to the regulation of transcription and translation, two important areas of current research in molecular biology.

Chapter 4

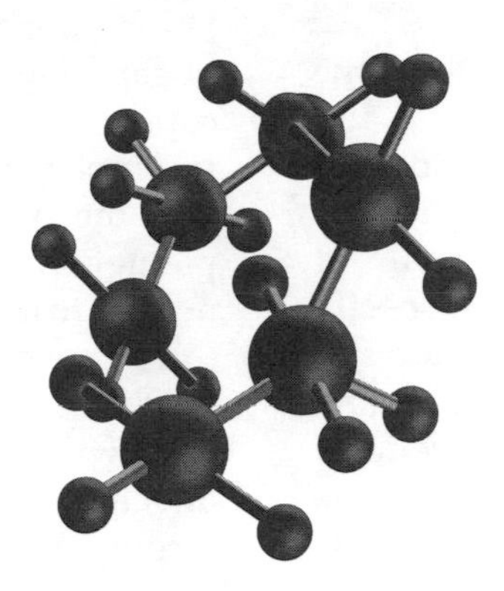

Transcription and Its Control

Ribonucleic acid or RNA results when a section of DNA is copied in the nucleus during transcription. RNA plays an important role in the conversion of the DNA information into a sequence of amino acids in a protein (remember, proteins are the action molecules of an organism).

Types of RNA

Messenger RNA (mRNA) is the actual template for protein synthesis in the cytosol. This means that the base sequence on mRNA specifies the sequence of amino acids in a polypeptide chain. Most genes will generate mRNA, the lifetime of which is short, usually several hours or days. Messenger RNA is also the least abundant of the three types of RNA.

Transfer RNA (tRNA) is the smallest of the RNA molecules, usually between 73 to 93 nucleotides (Nt) in length. Transfer RNA attaches to specific amino acids and brings them to the complex of mRNA and ribosomes on which a polypeptide is formed.

Ribosomal RNA (rRNA) is the most abundant and represents about 90% of all the RNA in a cell. A ribosome is a complex of protein and ribosomal RNA where proteins are synthesized. Figure 4.1 shows that the initial product of gene transcription must undergo modification to generate a specific kind of RNA molecule.

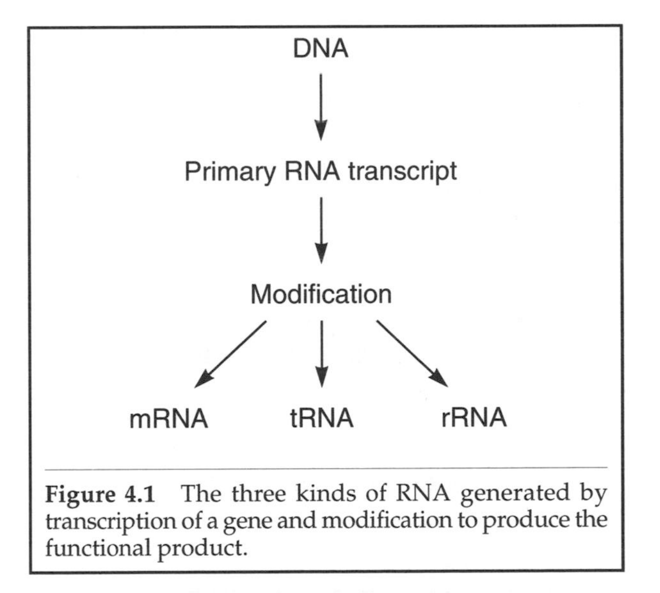

Figure 4.1 The three kinds of RNA generated by transcription of a gene and modification to produce the functional product.

The Genetic Code

The sequence of four bases in DNA is transcribed to form a four-base sequence in mRNA, which must specify the sequence of the 20 different amino acids used to make proteins. The only way that four different bases in mRNA can specify 20 different amino acids in a polypeptide chain is for the bases to be read in groups of three, known as codons. Four different bases read three at a time gives rise to 4^3 different possibilities, or 64 codons. This process is called triplet coding, that is, three bases read together to give a message.

The genetic code is the relationship between the base sequence of DNA, transcribed to mRNA, and the sequence of amino acids in a polypeptide. We usually consider the genetic code from the perspective of the codons of mRNA that spell out amino acids. Of the 64 possible codons, 61 specify amino acids, and the remaining three are stop signals. Because there are 61 codons for only 20 amino acids, most amino acids have two or more codons. The codons for an amino acid with more than one codon are very similar. This similarity makes sense if minor errors are made. For example, the four codons for the amino acid glycine are GGA, GGG, GGC, and GGU. Notice that these all contain the same first two bases (letters). If amino acids are similar in structure, their codons are similar. For example, aspartic acid has the codons GAC and GAU, whereas glutamic acid, which is closely related in structure, has the codons GAA and GAG. If there is a base reading error, the same or a similar amino acid will still result.

The codon AUG is the initiation codon that signals the start of translation. This codon also represents the amino acid methionine; thus the first amino acid used in protein synthesis is always methionine. However, not all functional proteins have methionine as the first amino acid, for it can be removed after the polypeptide is completely formed. The codons UAG, UGA, and UAA are stop codons; they say that translation of the mRNA message is ended.

The genetic code is universal for all organisms studied except in mitochondria, where a different code operates. However, the amount of mitochondrial DNA is small—16,569 base pairs in humans—just enough to code for 13 polypeptides, 2 kinds of rRNA, and 22 tRNA molecules.

Transcription

During transcription, a small section of one strand of the huge, double-stranded DNA molecule is copied to yield an RNA molecule. The RNA formed is complementary to the DNA strand that is copied, with the exception that in RNA the base U is complementary to base A in DNA. The precursors needed to make RNA are nucleoside triphosphates (NTP), such as CTP, GTP, ATP, and UTP. The strand of DNA copied is read in the 3′ to 5′ direction. The RNA will be formed in the 5′ to 3′ direction. These directions are the same as those involved in making a complementary copy of DNA when it is replicated prior to cell division.

The formation of RNA during transcription is catalyzed by an oligomeric enzyme known as DNA-directed RNA polymerase, or simply RNA polymerase, of which there are three kinds. RNA polymerase I (abbreviated RNAP I or sometimes Pol I) transcribes the 280 copies of the gene that generates most of the rRNA in humans. Pol I is the most abundant RNA polymerase. RNA polymerase II (RNAP II or Pol II) transcribes genes containing the information for the amino acid sequence in a polypeptide chain. This form of RNA polymerase gives rise to mRNA. RNA polymerase III (RNAP III or Pol III) is the least abundant of the RNA polymerase enzymes. Its products are tRNA and other small RNA molecules, one of which will form part of the ribosome. In the human genome, about 2,000 copies of this small rRNA gene are transcribed by RNAP III.

Figure 4.2 shows in simplified form what happens during transcription. Remember, there are two DNA polynucleotide chains. Part of one will be transcribed, producing a complementary copy in the form of the RNA molecule. Also remember, RNA contains uracil, not thymine. The DNA strand that is copied is the template strand. The other is the sense

	← Upstream direction	Downstream direction →
Sense strand	5′...A C G G T A A T G G C...3′	
Template strand	3′...T G C C A T T A C C G...5′	
RNA strand	5′...A C G G U A A U G G C...3′	

Figure 4.2 A section of double-stranded DNA showing the strand copied during transcription (template strand) and the untranscribed or sense strand. The RNA strand is complementary to the template and has the same base sequence as the sense strand with uracil (U) in RNA replacing thymine (T) in DNA.

strand because it will have the same base sequence as the RNA, except U will replace T. The polarity of the sense strand and RNA are the same and opposite to that of the template strand. Directions are often described in terms of river flow. Upstream refers to the 5′ direction and by convention the frame of reference is the sense strand of DNA. Likewise, downstream refers to the 3′ direction of flow.

Figure 4.3 shows the three major phases of transcription. Initiation begins when general transcription factors (proteins, which we will discuss next) and RNA polymerase bind to the double-stranded DNA just upstream of the start site, forming a preinitiation complex. The start site is where transcription actually begins, and the first DNA base copied is given the number +1. Bases just upstream of this are in a region called the promoter and are given negative numbers. When RNA polymerase and the general transcription factors bind to the promoter region, the DNA at the start site is unwound, exposing the template strand that will be copied. The first nucleotide triphosphate comes in, recognizing and binding to its complementary base on the template strand via hydrogen bonds. Then the second NTP comes in, recognizing its complementary base, number 2 on the template strand. A phosphoester bond is formed between the 3′OH of ribonucleotide one and the 5′ phosphate of ribonucleotide two, and a pyrophosphate is released. During initiation, the RNA polymerase does not move along the DNA.

In the elongation phase, the RNA polymerase moves along the template strand of DNA, making

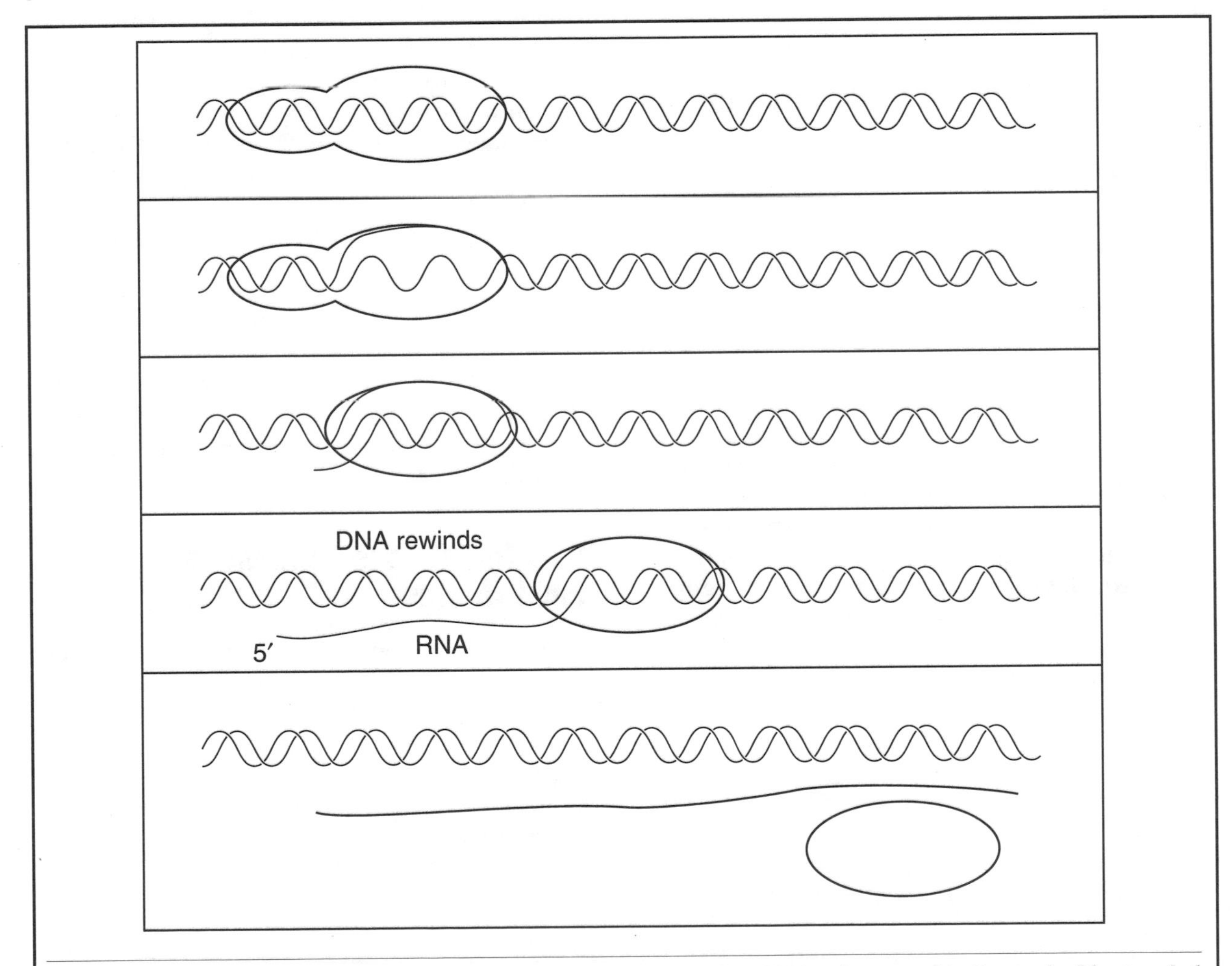

Figure 4.3 Transcription occurs in three distinct stages. Initiation involves RNA polymerase binding to double-stranded DNA along with accessory proteins, shown as an extension to the polymerase enzyme (top panel). DNA is unwound, exposing the template strand to be copied (second panel). During elongation, the polymerase moves along the DNA, making a strand of RNA complementary to the DNA template by adding one base at a time (third panel). The DNA is unwound as the polymerase moves along, whereas behind the enzyme DNA rewinds (fourth panel). When a sequence of bases spelling out a termination message is reached, the RNA polymerase and the primary RNA transcript molecule dissociate from the DNA (bottom panel).

a complementary RNA strand. As it moves, it unwinds the DNA double helix, catalyzing the formation of the RNA strand. Behind it, the DNA double helix re-forms. The template strand is read in the 3′ to 5′ direction, and the RNA strand is formed in the 5′ to 3′ direction. When the bond forms between the terminal free 3′OH of the growing RNA polynucleotide and the incoming NTP, only one of the three phosphate groups is needed; the other two are released as inorganic pyrophosphate, PPi—the same as occurs in DNA replication.

The termination phase begins when the RNA polymerase has moved along the template strand of DNA and reaches a sequence of bases that indicates the gene message is terminated. At this point, the RNA polymerase and the RNA strand dissociate from the DNA.

Regulation of Transcription

Each cell nucleus contains all the genes for that organism as base sequences in the DNA. However, cells only express the genes they need. For example, a liver cell will express genes also expressed by other cell types, but it will also express a number of different genes compared, for example, to a nerve or heart muscle cell. The human body contains more than 250 different kinds of cells. Also, as cells differentiate during development, different genes are expressed, giving rise to new proteins, and formerly active genes are no longer expressed. Furthermore, adult cells adapt to new circumstances by expressing previously inactive genes and not expressing previously active ones. Clearly, transcription must be regulated. This means that it is necessary to regulate what genes are expressed and at what rate. Although every human cell nucleus contains about 100,000 genes, only a fraction of these are transcribed at any one time, and those that are expressed depend on the cell type and its stage of development.

In terms of gene expression, two major types of control decisions must be described for a cell. Irreversible decisions turn specific genes on or off completely. For example, during embryo development, certain genes are initially expressed then turned off completely. Other genes are irreversibly turned on. But the second type of decision *is* adjustable, in terms of transient increases or decreases in the rate of transcription of an already expressed gene in response to various environmental or metabolic conditions. Consider the analogy of a light switch with a dimmer. The irreversible decision is that the switch is either on or off. Once on, however, it can be adjusted to produce a low, moderate, or high level of light intensity.

Most Eukaryotic Genes Are in Pieces

In eukaryotic organisms (organisms containing a nucleus) the genes are interrupted, or put another way, in pieces. Parts of the gene, known as exons, contain information that will appear in mature mRNA. Other parts contain base sequences, known as introns or intervening sequences, that will not appear in the mature mRNA (see Figure 4.4). Both introns and exons are still DNA, and although they appear to be different, there is no apparent structural difference between them. Note in Figure 4.1 that, after the transcription process, the modification step involves removing the introns from the primary RNA transcript. What is left are the exons, whose base sequences in mRNA will produce amino acid sequence information for a polypeptide. With a size of 2.4 Mb, the gene for Duchene muscular dystrophy is the largest in the human genome. Its introns are as long as 100 to 200 kb, and it may take 24 hours to be transcribed.

Cis, Trans, Base Sequences, and Proteins

Proteins are responsible for most of what happens in an organism, including the regulation of gene transcription. Regulatory proteins are called transcription factors. They are products of their own genes and most likely on different DNA molecules from the gene or genes they regulate. They are also called transactivating or trans factors. If they promote transcription, they are activators, and if they have a negative effect on transcription, we call

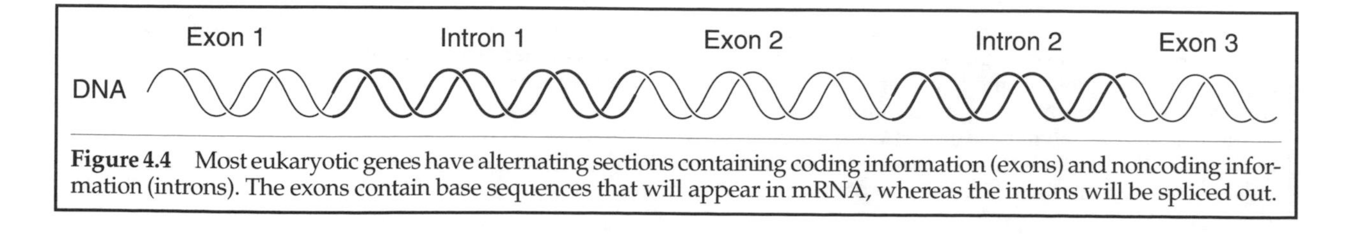

Figure 4.4 Most eukaryotic genes have alternating sections containing coding information (exons) and noncoding information (introns). The exons contain base sequences that will appear in mRNA, whereas the introns will be spliced out.

them repressors. We can define two major classes of trans factors: general factors needed for initiating transcription from most protein-coding genes and those that modulate general transcription factors and provide specificity for cell types and differentiation.

To regulate transcription, trans factors must bind to specific DNA sequences known as cis-regulatory elements, or cis elements, because they are on the same DNA molecule as the gene they regulate, usually very close to the base sequence being transcribed. A cis-regulatory element is a base sequence described by any one of the following terms: a motif, an element, or a box. They are normally short, usually 6 to 10 bp (base pairs). These motifs, boxes, or elements have a negative charge because of the sugar phosphate backbone of the DNA strands. They are found in a region 100 or so base pairs upstream of the start site of transcription but can be located much farther away. The region just upstream of the transcription start site is called the promoter region. Other cis elements are often farther upstream of the transcription start site, downstream from the gene itself, or within the gene, perhaps in an intron. They are known as enhancers if they act to promote transcription, or silencers if they hinder transcription. By convention, cis elements are described in terms of the sense strand; the term upstream describes the 5′ direction from the transcription start site, and downstream describes the 3′ direction.

The trans factors (regulatory proteins) do not work in isolation. They must obviously bind to specific cis motifs (on DNA) and to other trans factors because more than one trans factor is always involved in gene transcription. Therefore, trans factors must have amino acid sequences that both recognize and bind to a specific cis motif on DNA as well as amino acids that bind other trans factors to carry out their activating role. Accordingly, the amino acid region of the trans factors that binds to the cis motif must have a positive charge to interact with the negatively charged DNA phosphate groups. The region of the trans factors that recognizes and binds to other trans factors in activation must have a specific recognition conformation. A number of protein-DNA and protein-protein binding regions have been identified and given interesting names, such as the leucine zipper (every seventh amino acid is the hydrophobic amino acid leucine), zinc fingers (contains the metal ion zinc), and the helix turn helix (a short region of α helix, followed by a short looping section, then another region of α helix).

The General Transcription Apparatus

Nearly all genes coding for proteins in eukaryotes have a TATA box in their promoter region. The TATA box is an adenine-thymine (AT) only 8-base sequence found 25 to 30 bp upstream of the transcription start site, where a transcription preinitiation complex is formed, involving general transcription factors and RNAP II (Pol II). This site is likely involved in properly positioning RNAP II upstream of the transcription start site so proper initiation of transcription occurs.

The general transcription factors include POL II (we are sticking with genes for proteins, hence the use of II) and about seven other proteins identified as TFIIA, TFIIB, TFIID, TFIIE, TFIIF, TFIIH, and TFIIJ. Most are oligomeric proteins with a variety of subunits. TFIID, in particular, contains a number of subunits, including a TATA binding protein (TBP) subunit and several associated factors. Along with POL II, the general transcription factors establish a basal level of transcription. Modulation of this rate as well as tissue- and time-specific activation involve additional transcription factors and cis elements to be discussed next.

The sequence of steps in the formation of a preinitiation complex on a promoter containing a TATA box is

1. binding of the TFIID to the TATA box;
2. binding of TFIIA and TFIIB to TFIID to form the DAB complex;
3. binding of RNAP II along with TFIIF to the DAB complex;
4. binding of TFIIE; and
5. binding of TFIIH, followed by TFIIJ to complete the preinitiation complex.

Figure 4.5 shows the binding of all the general transcription factors at the TATA box. Transcription factors are drawn as balloons, and subunits are not shown.

Induced Levels of Transcription

In addition to the TATA box which is so widely found in the promoter region of protein-coding genes, there are other base sequences recognized by specific proteins that can lead to higher levels of transcriptional control. These cis elements are enhancers or silencers, and they may be close to or far removed from the TATA box along the DNA molecule. Remember, transcription factor proteins

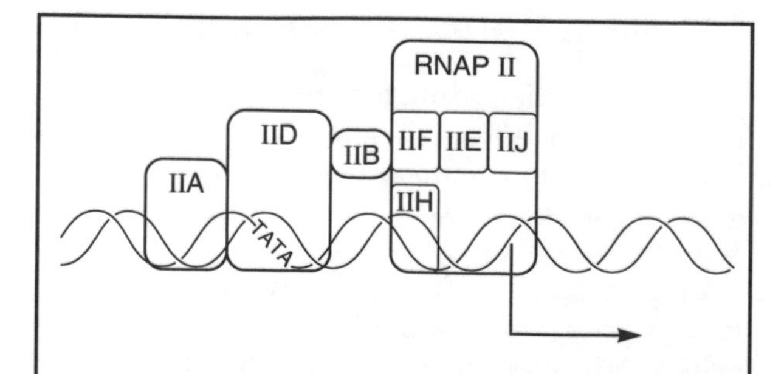

Figure 4.5 A schematic to show the final assembly of the general transcription complex necessary for proper transcription of genes and containing a thymine (T)-adenine (A)-TATA box in the promoter. The individual transcription factors are shown as balloons, each identified by II and a letter. RNA polymerase II (RNAP II) is shown near the transcription start site, indicated by the bent arrow. Both strands of DNA are shown, and the only bases identified are those in the TATA box where transcription factor IID binds.

that bind to cis elements are activators or activator proteins; if they inhibit transcription they are repressors. The trans factors that bind to these cis elements interact with the general transcription factors to modulate transcription of specific genes. This interaction is achieved in the preinitiation complex by bending or looping of DNA so distal cis elements with their bound trans factors can interact with the general transcription factors at the point where RNAP II binds just upstream of the transcription start point. Figure 4.6 illustrates how DNA looping brings distal transcription factors into contact with the general transcription factors at the start site of transcription to regulate the overall transcription of a gene.

The trans factors that augment or inhibit transcription at the point of initiation may be ubiquitous, that is, found in a variety of cell types, or highly specific. Four highly specific trans factors, found only in skeletal muscle and essential for their development, are MyoD, myogenin, MRF4, and myf5, which contain secondary and tertiary structural elements that recognize a six-base sequence (cis element) known as an E-box. Artificial expression of the MyoD gene family in other cell types can result in those cells expressing genes only transcribed in skeletal muscle.

A variety of hormones and related molecules act within a variety of cell types by directly regulating gene expression. The steroid hormones (glucocorticoids, testosterone, estrogen, and progesterone), thyroid hormone, the active forms of vitamin D, and retinoids (from vitamin A) circulate in the blood and readily enter cells because their lipophilic nature permits them to diffuse across the cell membrane. Inside the cell, they bind to protein receptors which are themselves products of specific genes. When the hormone (or ligand) is bound to its receptor, the receptor conformation is structurally altered.

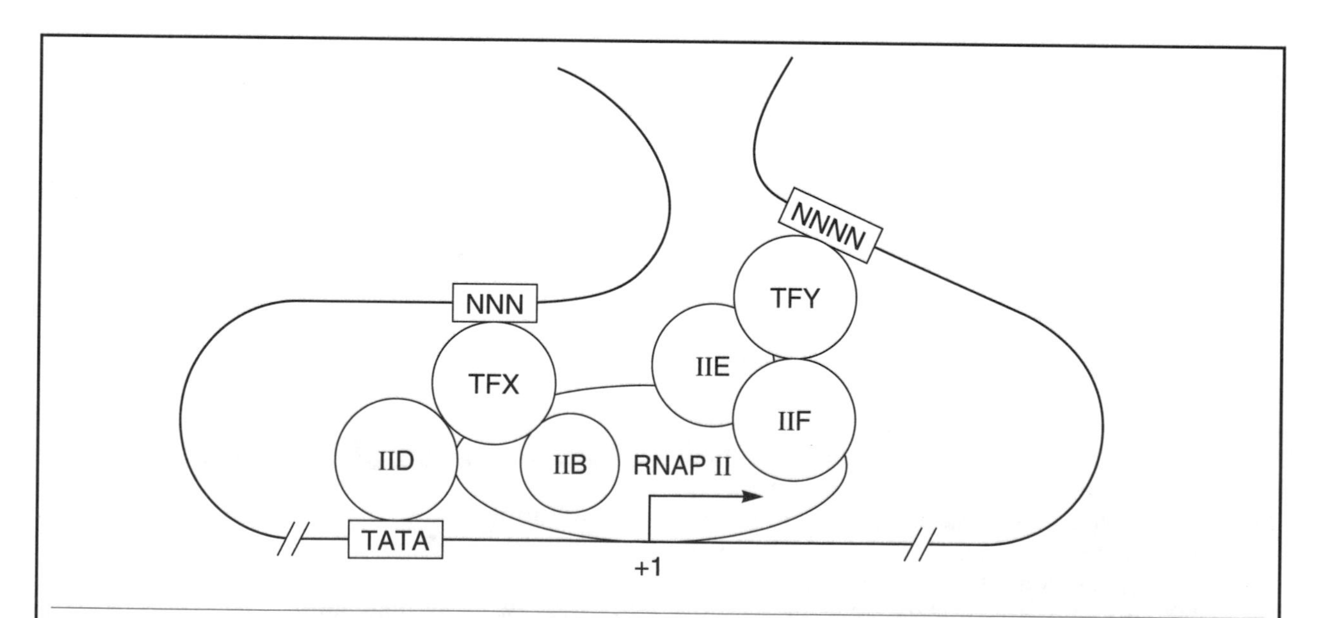

Figure 4.6 A model to illustrate how looping of DNA allows distal cis elements, identified by boxes and the letter N to indicate unspecified bases, to interact through their transcription factors (TFX and TFY) with the general transcription factors (IIB, D, E, & F) at the site of the initiation. Only some of the general transcription factors are shown, along with RNA polymerase II (RNAP II). Only a single strand of DNA is shown. The transcription start site is indicated by the bent arrow and the +1.

The hormone-receptor complex can now bind to specific cis elements, known as hormone response elements (HRE), in the vicinity of target genes to regulate their expression. Gene regulation occurs when two hormone receptors with bound ligand bind at the HRE. This binding is a common feature of specific gene activation. If the two transcription factors (or, in this case, hormone receptor plus hormone) are the same, they are homodimers; if they are different, they are heterodimers. Figure 4.7 illustrates a common sequence for steroid hormone activation of a gene.

Often, HREs are located upstream of the TATA box, but some, like enhancers, are also several thousand base pairs from the start of transcriptional initiation. The ligand-receptor complex, bound to its specific HRE, helps induce the formation of a transcription initiation complex, and along with RNAP II, promotes the beginning of transcription. Binding of the ligand-receptor complex to the HRE may promote a conformational change in DNA, opening the gene for transcription or repressing the gene by competitively blocking access of other transcription factors near the transcription start site.

In Figure 4.6, transcription factors X and Y (i.e., TFX and TFY) could represent hormone receptor complexes, bound at their specific HREs and interacting with the general transcription complex. Artificial analogues of testosterone, such as anabolic steroids, can stimulate skeletal muscle growth, likely by increasing the transcription of certain muscle protein genes.

Exercise training can also increase the transcription of genes that code for enzymes involved in mitochondrial metabolism. These genes are usually located in nuclear DNA, but genes on mitochondrial DNA are also transcribed more rapidly with endurance exercise. Evidence suggests that genes coding for enzymes involved in glycolysis are transcribed at a slower rate with aerobic training. Whether regularly performed exercise alters the expression of genes coding for contractile proteins is also being studied. For example, the gene for the IIB heavy chain of myosin may be transcribed less and the IIA myosin heavy-chain gene expressed more during short-term endurance training. However, an actual change in the transcription of the type I myosin heavy-chain gene has not been adequately demonstrated in human exercise training studies.

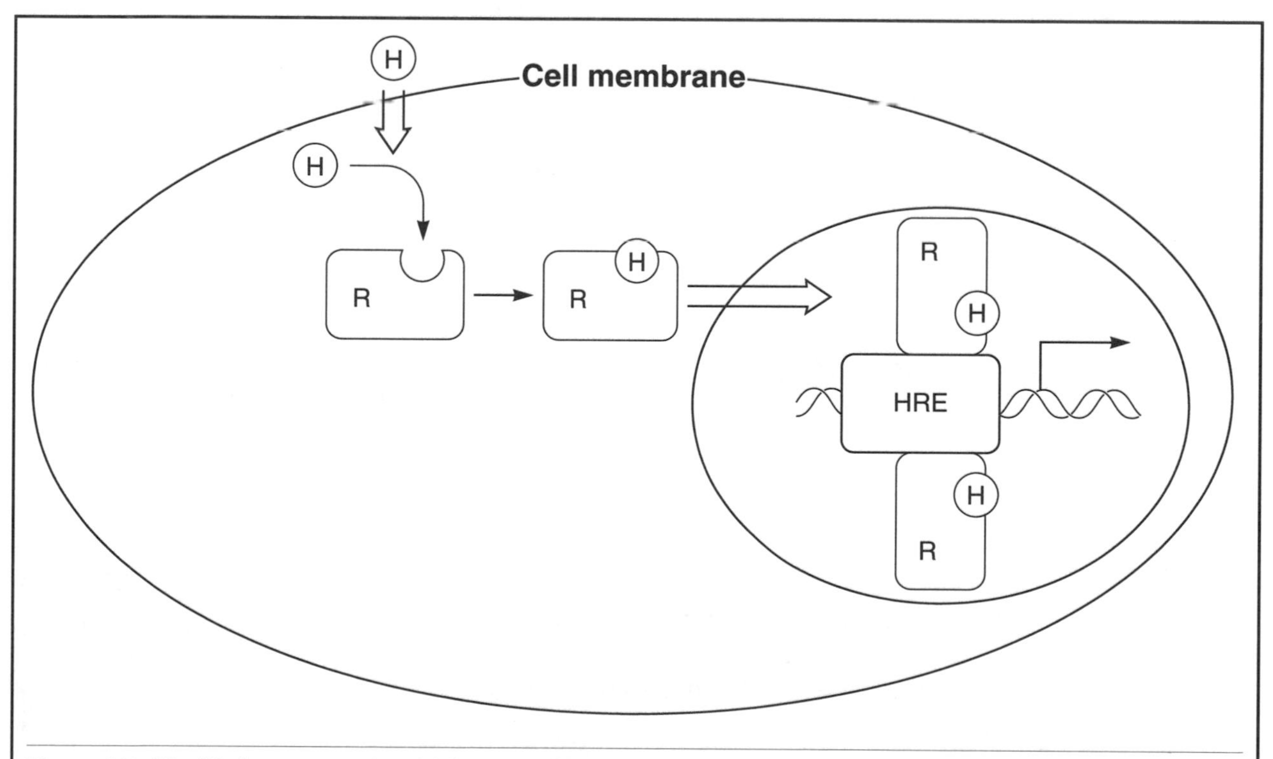

Figure 4.7 The likely sequence in which a steroid hormone (H) influences gene transcription. The hormone, being lipophilic, diffuses across the cell membrane and interacts with its specific receptor (R). The hormone receptor complex now crosses the nuclear membrane and interacts with its specific hormone response element (HRE) in the promoter region of a gene. Pairs of hormone receptor complexes bind as homodimers, along with the other members of the general transcription complex (not shown) to increase (or decrease) the rate of transcription initiation from the start site, shown by a bent arrow.

Many transcription factors and the specific cis elements in DNA to which they bind have been identified. Figure 4.8 illustrates different trans factors, indicated by balloons and their common acronyms, and their cis elements. For example, the thyroid hormone receptor (THR) with bound thyroid hormone binds to the cis element of base sequence AGGTCATCACCT, known as the thyroid hormone response element or TRE. The skeletal muscle specific transcription factor, MyoD, also binds to its specific base sequence or E box. Because many different specific transcription factors must interact with the general transcription complex and with other specific transcription factors, there are wide variations in the level of control of specific genes. Furthermore, specific cell types express their own transcription factors as well as those found in other cells and in the general transcription complex (i.e., TFIID, TFIIE, etc.). Even though a steroid hormone enters a cell, it cannot influence transcription unless that cell expresses the gene for that hormone receptor. Thus, by controlling expression of transcription factor genes or genes for steroid hormone receptors, cells and tissues maintain their unique character.

The presence of some trans factors (i.e., whether their genes are expressed) depends on signals originating outside the cell, such as nerve stimuli, growth factors, or hormones. Not all hormones enter the cell. Many hormones and all growth factors interact with specific receptors on the cell membrane and generate internal signals that influence gene expression. Hormones, growth factors, and nerve signals allow the multi-celled organism to control events within individual cells so that the organism operates in an integrated, cohesive manner.

Regulation by Organization of DNA

Although each nucleus contains the same DNA molecules in each chromosome, differences exist in the actual structural organization of the DNA.

The genes of a chromosome in an active or open conformation can be transcribed. In another cell, these genes may be repressed simply because the promoter region is closed to transcription factors by the way the DNA is coiled within the chromosome. Thus the organization of genetic information in a cell can determine what genes are expressed.

Another mechanism to regulate gene expression is the modification of certain bases in the promoter region so that they are not recognized by transcription factors. Certain cytosine bases next to guanine in promoter regions of eukaryotic DNA can be altered by the addition of a methyl group (CH_3). Methylation in upstream regulatory sites is associated with transcription inactivity. Methylation and its reversal, demethylation, may also regulate the expression of certain genes in transcriptionally active areas, providing a further level of transcription control.

Clearly, regulation of transcription is an important and complex event in a cell, involving multiple levels of control, such as nerve signals, hormones, and growth factors; the activity of the cell; and the nutritional state of the organism. Moreover, the DNA arrangement in specific cell types can determine which genes are spatially accessible for activators to bind to them or their upstream and downstream regulatory base sequences. Failures in the normal expression of genes can underlie many diseases, such as cancer.

Before leaving this topic, let us look at housekeeping genes, or those constitutively expressed in a wide variety of cell types at a constant rate with little or no fluctuations. These genes generally code for proteins that function as enzymes, such as those involved in glycolysis. These genes normally lack a TATA box and are under less complex control by cis elements and their trans factors. Areas rich in GC doublets are typically found in the upstream regions of these genes.

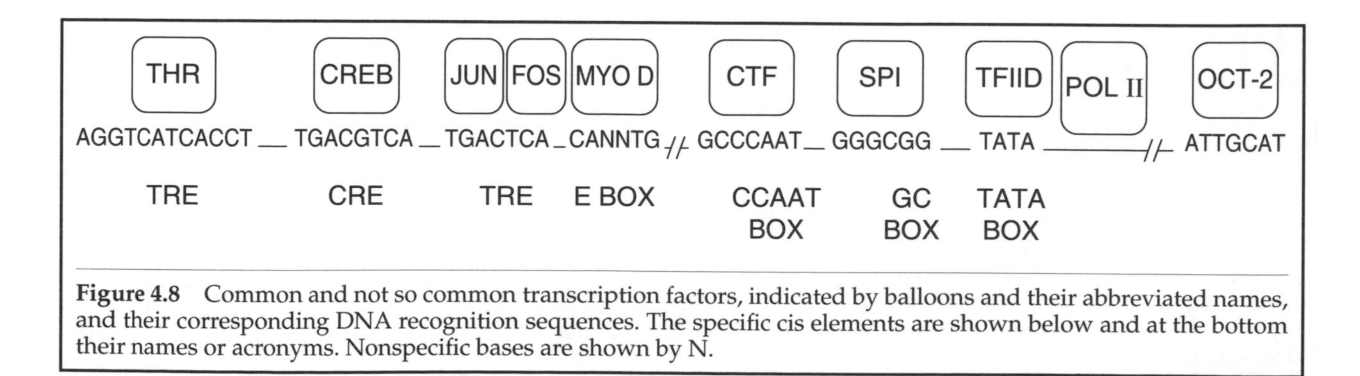

Figure 4.8 Common and not so common transcription factors, indicated by balloons and their abbreviated names, and their corresponding DNA recognition sequences. The specific cis elements are shown below and at the bottom their names or acronyms. Nonspecific bases are shown by N.

Summary

Transcription of genes produces an RNA molecule called the primary RNA transcript. Depending on the gene transcribed, the transcript will be modified to produce a messenger RNA (mRNA), transfer RNA (tRNA), or ribosomal RNA (rRNA). Modification of the primary transcript into mRNA involves adding a cap at the 5′ end, adding a series of 200 to 250 adenine nucleotides to the 3′ end, and removing base sequences (introns) that do not provide amino acid sequence information in the protein product. The base sequence of the mRNA is read in groups of three, called codons, which constitute the words of the genetic code. Sixty-one of the possible 64 codon combinations spell out the 20 amino acids, so the genetic code is said to be degenerate. Three of the codons (UAA, UGA, and UAG) stop the message. Transcription is carried out by a group of enzymes known as RNA polymerases. These enzymes copy the genes that will become rRNA (RNAP I or Pol I), mRNA (RNAP II or Pol II), or tRNA (RNAP III or Pol III). During transcription, the polymerase unwinds the two DNA strands and copies part of one, reading it in the 3′ to 5′ direction. The nontranscribed DNA strand, called the sense strand, has the same base sequence as the primary transcript, except that thymine, in DNA, is substituted by uracil in RNA.

In the region before the transcription start site, in a direction described as upstream, specific protein molecules recognize base sequences to regulate the process of transcription. The base sequences, or promoters, are known as cis elements, and the proteins that bind are called trans factors. Two types of trans factors are identified. General transcription (trans) factors are needed for the transcription of all genes. These bind upstream of the start site at a cis element called the TATA box and help position RNA polymerase and start the transcription process. Virtually all genes require other tissue or developmentally specific proteins to provide an additional level of transcription control. They may be activators or repressors that bind to cis elements known as enhancers or silencers, respectively. Circulating hormones and growth factors can influence transcription by entering the cell, binding to specific receptors, and modulating the process; members of the steroid hormone family work this way. Other hormones and factors bind to receptors on the cell membrane and generate messenger molecules within the cell that alter the rate of transcription.

Chapter 5

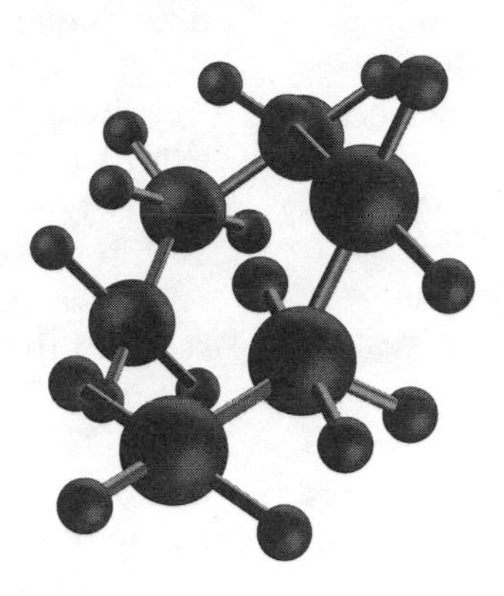

Protein Synthesis

Proteins are synthesized on ribosomes in the cell cytosol. The messenger RNA (mRNA) message is translated into a sequence of amino acids. Before any of this can take place, however, the RNA molecules made by transcription must be modified to generate the active forms of transfer RNA (tRNA), ribosomal RNA (rRNA), and mRNA.

Posttranscriptional Modifications of RNA

Most of the expressed gene are single copy genes and provide information for the sequence of amino acids in a polypeptide. However, there are multiple copies of genes that provide the information for the formation of RNA molecules destined to become rRNA and tRNA.

Formation of mRNA Molecules

As you will recall, transcription of gene coding for polypeptides (proteins) or for rRNA or tRNA proceeds in the 5′ to 3′ direction. Before the transcription process is very far along, a cap is added to the 5′ end. This involves adding a type of guanine nucleotide, the details of which are not important here. Then, a poly A tail is added to the completed 3′ end for most RNA molecules destined to become mature mRNA. The poly A tail consists of about 200 to 250 adenine nucleotides, added by the action of an enzyme known as poly A polymerase. In mammals, this enzyme begins adding the poly A tail 20 to 30 nucleotides after the base sequence AAUAAA on the RNA, at a point where a GU-rich sequence begins. In the sense strand of DNA, the sequence would be AATAAA, and the RNA POL II would pause at a GT-rich sequence. The poly A tail is not part of the DNA transcription process but is added afterwards.

The next step is to remove the introns from the capped, polyadenylated, primary gene transcript via a splicing process in which the junction between introns and exons is cleaved at both the 5′ and 3′ ends. The introns are then removed and contiguous exons joined up. The details of this process are complicated, involving small nuclear RNA molecules known as snRNAs. The snRNAs are associated with proteins to create a complex known as small nuclear ribonucleoproteins, or "snurps."

As a result of the 5′ capping, the 3′ polyadenylation, and intron splicing, we now have what has been called a mature mRNA, which then leaves the nucleus for the cytosol where it acts as a template on which a polypeptide (protein) will be made employing ribosomes. We will look more closely at this process later.

It is not uncommon for one gene to give rise to two or more mRNA molecules, and thus two or more final proteins. This characteristic is common for some of the contractile proteins in skeletal, cardiac, and smooth muscle and can occur in these instances:

1. Transcription is initiated at different promoters, resulting in different 5′ exons in the mRNA.
2. Transcription terminates differently because of more than one site of polyadenylation, resulting in different 3′ exons.
3. Different internal exons in the gene are included or not included, giving rise to different mRNA molecules (thus two or more different polypeptides) in a process known as alternate mRNA splicing. This alternative splicing can occur in the same cell at the same time, in the same cell at different times during cell differentiation, or in different cells. Control of alternative splicing occurs through the action of specialized proteins.

Formation of rRNA and tRNA Molecules

Modification of primary transcripts from genes for rRNA and tRNA also occurs. As mentioned, two different genes are needed to produce the rRNA molecules that make up much of the ribosomes. The products of these genes or primary transcripts are described by their velocity of sedimentation in an ultracentrifuge, expressed in Svedberg units (S).

Note: *For very large molecules or molecular complexes, size is measured by how far they travel in an ultracentrifuge, expressed in Svedberg units (S) rather than molecular weight. The larger the molecule, the faster it sediments in a centrifuge, and the larger the value of S. Unlike units such as kilodaltons (kDa) or base pairs (bp), typically used to describe the size of protein or DNA molecules, respectively, Svedberg units are not linearly related to size. Thus a whole ribosome, with a mass of 80S, is made up of two subunits of size 60S and 40S because the sedimentation behavior of particles in the centrifuge is not linearly related to size.*

The two rRNA transcripts are described as 45S and 5S. The smaller transcript, 5S (120 bases), remains unchanged, whereas the larger transcript, 45S, undergoes some splitting and modification, resulting in three new rRNA molecules: 18S, 28S, and 5.8S. The 18S rRNA molecule combines with approximately 30 protein molecules to make the small ribosomal subunit, described as the 40S subunit. The 5.8S, the 28S, and the 5S rRNA molecules combine with about 47 protein molecules to make the large ribosomal subunit, known as the 60S subunit. During protein synthesis, the 40S and 60S ribosomal subunits combine to make the complete ribosome, described as the 80S ribosome.

Primary transcripts for genes for the various tRNAs undergo some modifications. For example, nucleotides are removed from the 5′ and 3′ ends, and, if present, an intron is removed. Then the 3′ end has a nucleotide sequence CCA added. Finally, some of the bases are modified. Figure 5.1 shows the main features of a fully functional tRNA molecule. The 5′ end has a 5′ phosphate group. The 3′ end has the sequence CCA with a free 3′ hydroxyl group to which the amino acid is attached. Notice the hydrogen bonding that helps maintain the rough cloverleaf structure, characteristic of the tRNA molecules.

Three bases at the bottom of tRNA represent the anticodon bases. These will correspond to the codon for the particular amino acid, except that they will be complementary. For example, if the codon is AAC, the anticodon will be UUG. Because of the anticodon on tRNA, there must be at least one tRNA molecule for each amino acid. However, there is a bit of flexibility in the anticodon-codon binding because the first base in the anticodon (the one at the 5′ end that pairs with the third base at the 3′ end in the codon) can often recognize two bases. For example, the codons CUA and CUG could be recognized by the anticodon GAU. This flexibility in codon-anticodon recognition is known as wobble.

Translation

In the process of translation, which takes place in the cytosol of the cell, the mRNA message is converted into a sequence of amino acids in a polypeptide chain.

Formation of Aminoacyl tRNA

Each amino acid has at least one tRNA to which it will be attached. Each tRNA will have an anticodon that will match one of the codons (possibly more due to wobble) for that amino acid, if it has more than one codon. Each amino acid will also have a specific enzyme to catalyze the joining of its α carboxyl group to the 3′OH group of the terminal adenosine of the tRNA using an ester bond. The enzyme that joins the amino acid to its tRNA is known as aminoacyl tRNA synthetase. It is the responsibility of each synthetase to match the correct amino acid and its tRNA, as shown in the following equation (p. 42).

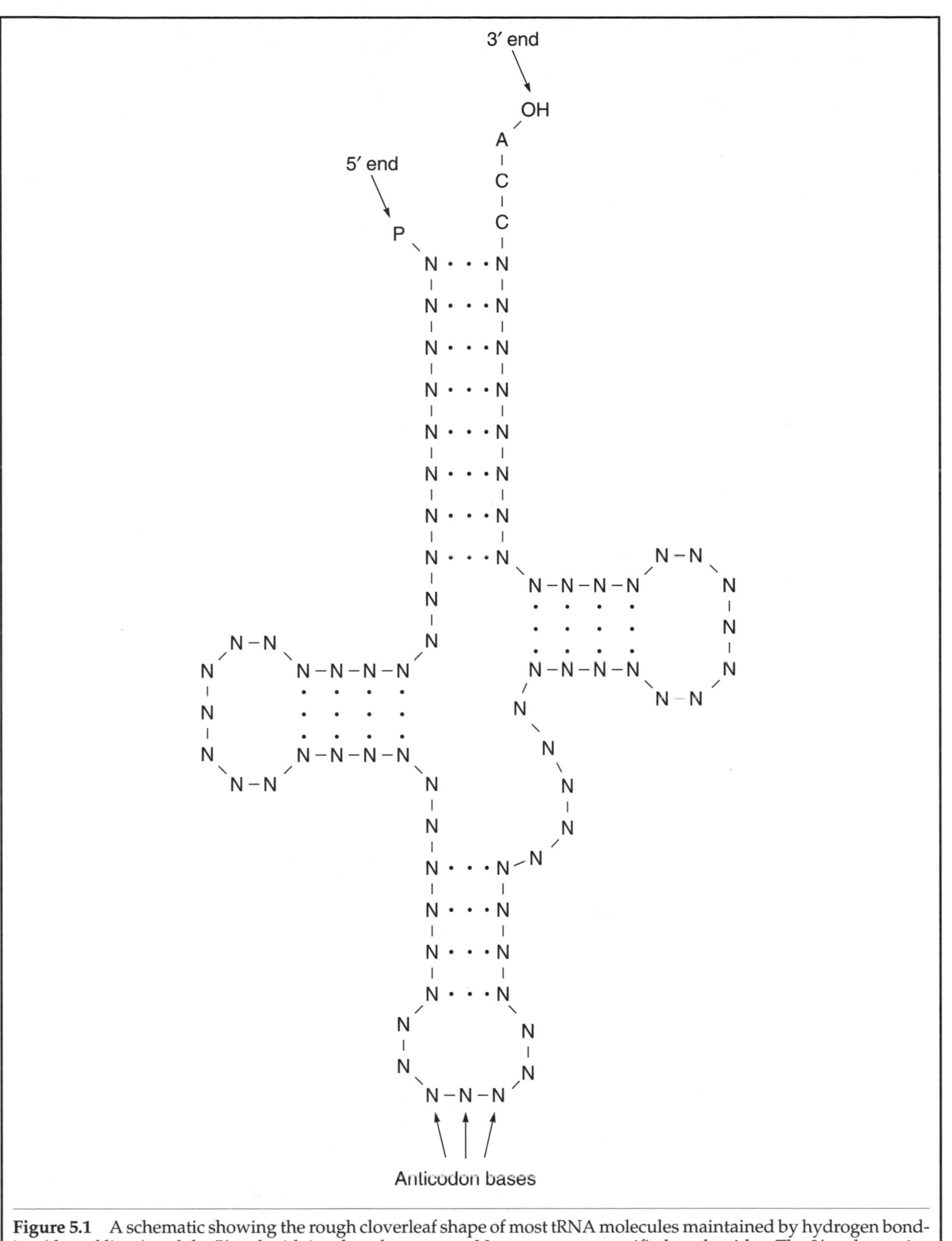

Figure 5.1 A schematic showing the rough cloverleaf shape of most tRNA molecules maintained by hydrogen bonding (dotted lines) and the 5′ end with its phosphate group. N represents unspecified nucleotides. The 3′ end contains the bases cytosine and adenine (CCA) and the 3′ OH (hydroxyl) group of the terminal adenosine is available to bind an amino acid. The three bases at the bottom represent the anticodon that recognizes the complementary bases on the mRNA codon.

$$\text{amino acid} + \text{tRNA} + \text{ATP} \xrightarrow{Mg^{2+}} \text{aminoacyl-tRNA} + \text{AMP} + \text{PPi}$$

The energy needed for the formation of the ester bond between the amino acid and its tRNA is provided by the hydrolysis of ATP. The products of this hydrolysis reaction are AMP and inorganic pyrophosphate (PPi). A magnesium ion is needed to make this reaction occur.

The Role of mRNA

Figure 5.2 illustrates the essential components of a functional mRNA molecule. The coding region, headed by the AUG start codon, contains the information to indicate the amino acid sequence for the polypeptide chain, which is followed by a stop codon (UAG, UGA, or UAA). Flanking the coding region is a 5′ noncoding region and a 3′ region that is also noncoding. Because the base sequence in these regions is not translated into an amino acid sequence, they are also known as the 5′UTR (untranslated region) and 3′UTR, respectively. Most mRNA molecules contain the poly A tail. The functional mRNA molecule, once formed in the nucleus, is transported to the cytosol where protein synthesis takes place.

Initiation of Translation

Like transcription, the translation process has three stages: initiation, elongation, and termination. We will not go into great detail in this section but will only focus on the overall process. The major players in the initiation of translation are the two ribosomal subunits, 40S and 60S, the mRNA molecule, the initial aminoacyl-tRNA (which will be methionyl-tRNA), a number of protein factors to control the initiation process (eukaryotic initiation factors such as eIF-1, eIF-2, etc.), and a source of energy from the hydrolysis of ATP and GTP.

Figure 5.3 illustrates the initiation part of translation. In the first step, the methionyl-tRNA (met-tRNA), an initiation factor, the 40S ribosome subunit, and GTP combine to form a preinitiation complex. Then mRNA binds to the preinitiation complex, forming the 40S initiation complex. During this step, the smaller ribosomal subunit (40S) binds to the functional mRNA, initially at the 5′ cap end, then moves along the mRNA in a 5′ to 3′ direction searching for the initiation codon AUG. The methionyl-tRNA will recognize the AUG (initiation) codon on mRNA because it is complementary to it. The 60S ribosomal subunit now binds, generating the 80S initiation complex. At least 10 initiation factors (eIFs) participate in the various stages of initiation. The hydrolysis of ATP and GTP are essential as energy sources for the initiation step.

Elongation of Translation

The elongation phase involves the addition of amino acids, one at a time, to the carboxy terminal end of the existing polypeptide chain (see Figure 5.4). Let us start with the 80S initiation complex. The next aminoacyl tRNA comes in, its anticodon recognizing the mRNA codon on the 3′ side of the initiator codon. A peptide bond is then formed between the methionine carboxyl and the free amino group of the next amino acid, still attached to its tRNA. The formation of this peptide bond means the methionine is released from its tRNA; the methionine tRNA leaves, and we now have a dipeptide attached to the second tRNA. The 80S ribosome complex then slides three bases along the mRNA molecule in the 3′ direction. Then the next aminoacyl tRNA comes in, recognizing the next mRNA codon. Another peptide

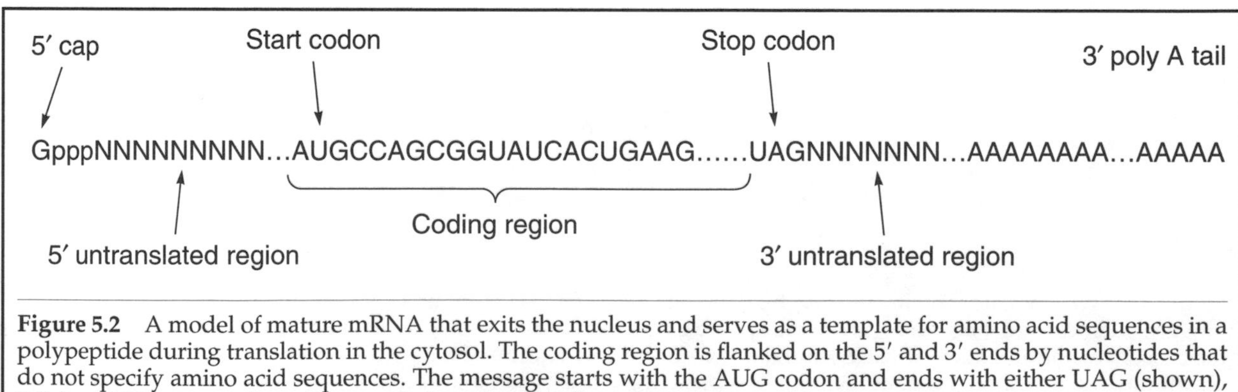

Figure 5.2 A model of mature mRNA that exits the nucleus and serves as a template for amino acid sequences in a polypeptide during translation in the cytosol. The coding region is flanked on the 5′ and 3′ ends by nucleotides that do not specify amino acid sequences. The message starts with the AUG codon and ends with either UAG (shown), UAA, or UGA stop codons. Unspecified nucleotides (bases) are shown as N.

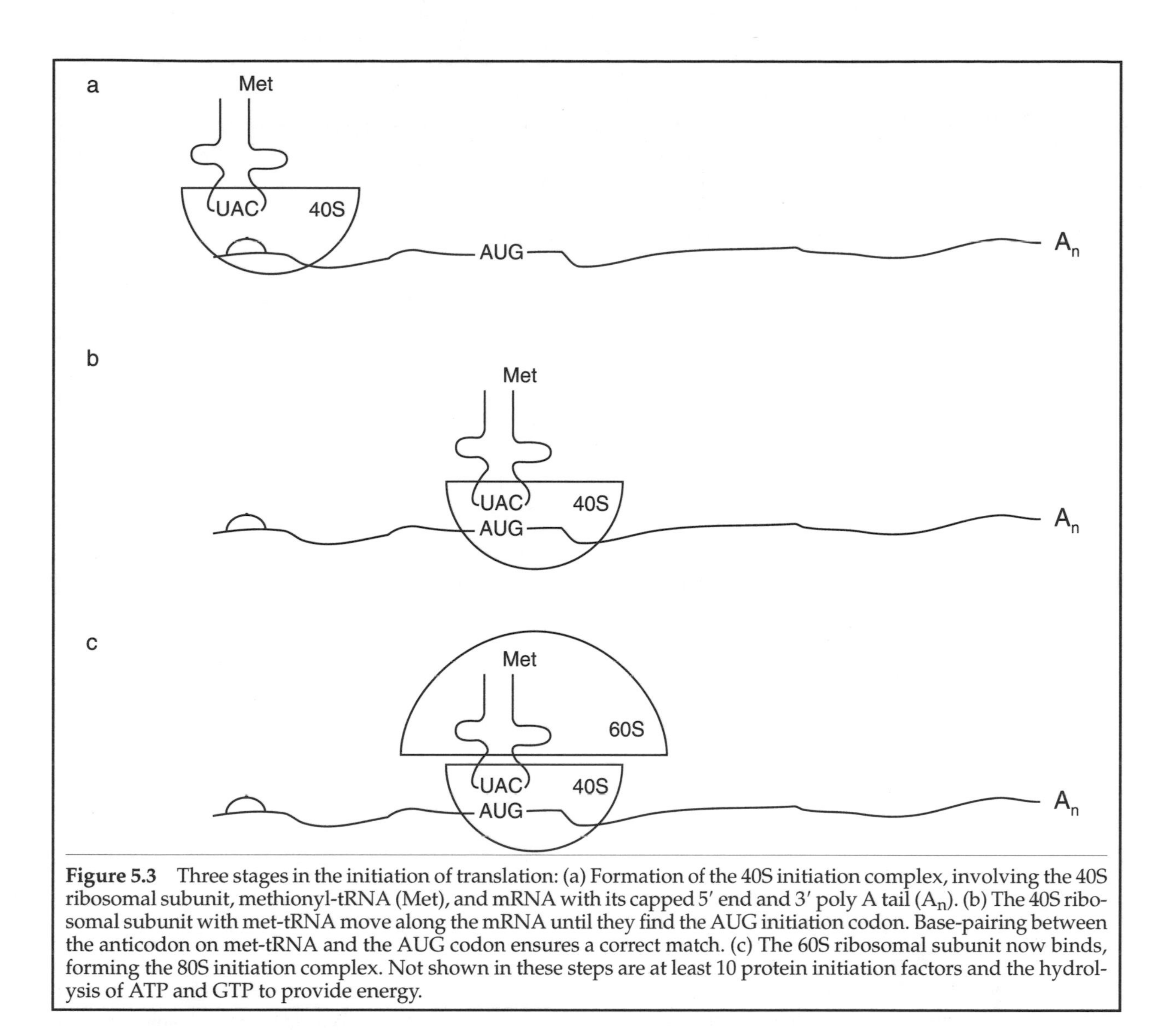

Figure 5.3 Three stages in the initiation of translation: (a) Formation of the 40S initiation complex, involving the 40S ribosomal subunit, methionyl-tRNA (Met), and mRNA with its capped 5′ end and 3′ poly A tail (A_n). (b) The 40S ribosomal subunit with met-tRNA move along the mRNA until they find the AUG initiation codon. Base-pairing between the anticodon on met-tRNA and the AUG codon ensures a correct match. (c) The 60S ribosomal subunit now binds, forming the 80S initiation complex. Not shown in these steps are at least 10 protein initiation factors and the hydrolysis of ATP and GTP to provide energy.

bond is formed, and we now have a tripeptide. Again, the 80S ribosomal complex slides along the mRNA. This process continues with the growing polypeptide chain still attached to the last incoming tRNA. The energy needed to make these peptide bonds comes from the hydrolysis of GTP to GDP, and Pi. Four proteins, called elongation factors (eEFs), are also needed.

Termination of Translation

The process of elongation continues, with the ribosome and growing polypeptide chain moving along the mRNA three bases at a time, until a stop codon is reached (UAA, UGA, or UAG). At this point, a termination factor causes the release of the completed polypeptide chain from the last tRNA. The whole complex dissociates, and the 80S ribosome dissociates into its two 40S and 60S subunits.

Because mRNA is large, more than one ribosome can be found on a single mRNA molecule—each containing a growing polypeptide chain. We call the mRNA and the two or more ribosomes with their growing polypeptide chains a polyribosome or polysome (see Figure 5.5).

Regulation of Translation

The process of translation can be rapidly altered in cells in response to specific stimuli or the nutritional state of the cell. This would suggest that the amount of mRNA acting as a template does not change because this would mean a change in the transcription rate, processing of the primary transcript in

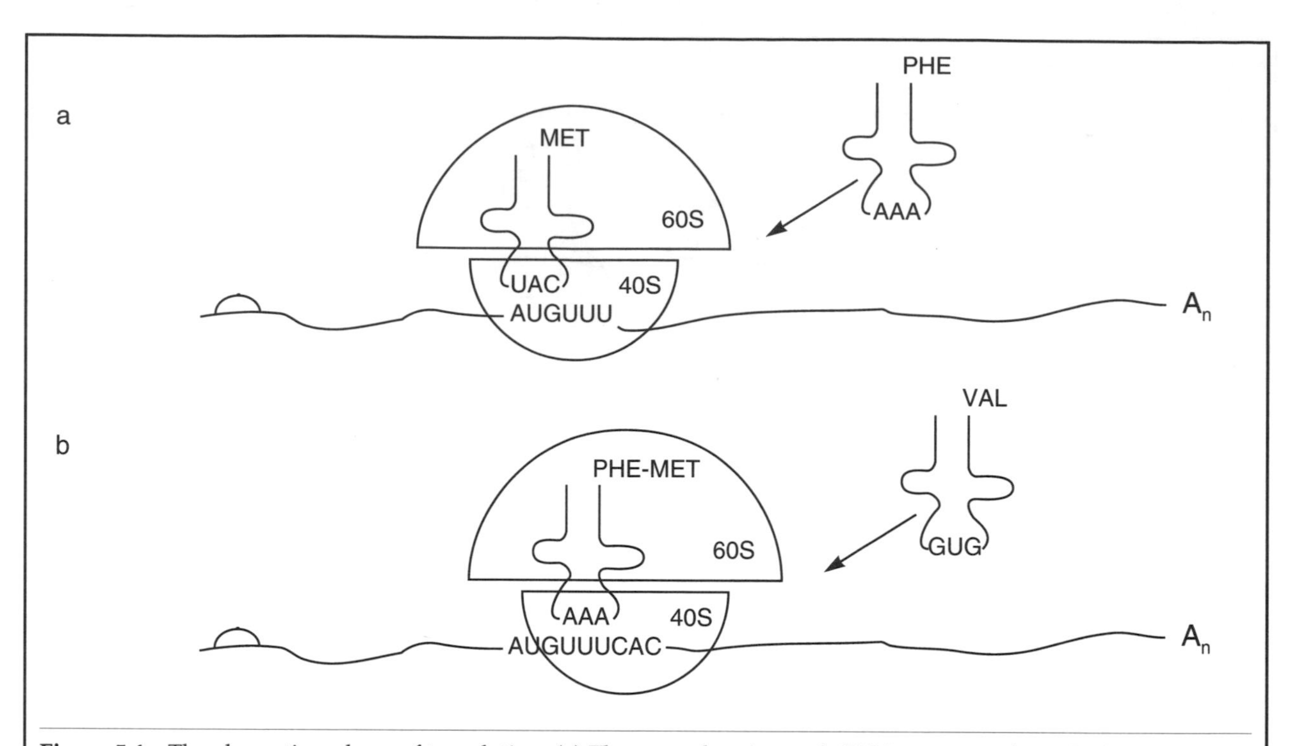

Figure 5.4 The elongation phase of translation: (a) The second aminoacyl-tRNA comes to the 80S ribosomal complex, its anticodon AAA recognizing the second codon UUU, which codes for the amino acid phenylalanine (PHE). (b) A peptide bond is formed between the carboxyl group of methionine (MET) and the phenylalanine, still attached to its tRNA. The 80S ribosome complex has moved three bases along the mRNA towards the 3′ end where the poly A tail (A_n) is located. The 5′ end contains the cap. Not shown are four elongation factors and the hydrolysis of GTP, which provides the energy for peptide bond formation and movement along the mRNA molecule.

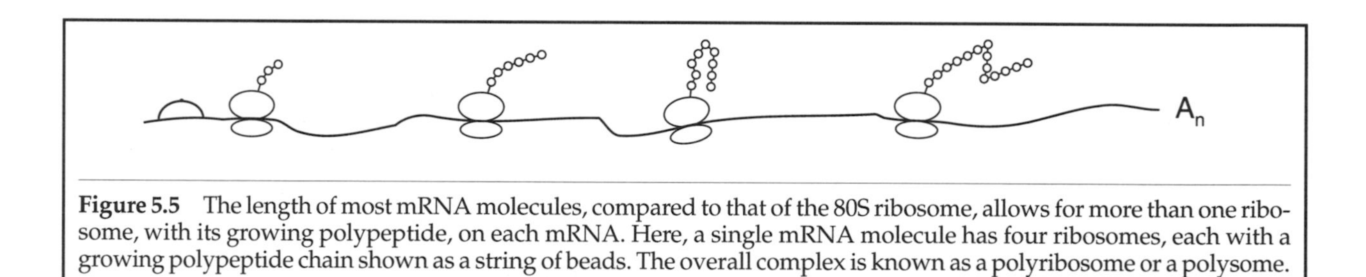

Figure 5.5 The length of most mRNA molecules, compared to that of the 80S ribosome, allows for more than one ribosome, with its growing polypeptide, on each mRNA. Here, a single mRNA molecule has four ribosomes, each with a growing polypeptide chain shown as a string of beads. The overall complex is known as a polyribosome or a polysome.

the nucleus to generate functional mRNA, and transfer of the mRNA to the cytosol—steps that take some time to complete. Research reveals that mRNA can exist in an inactive form in the cytosol, bound to proteins. In fact, about one third of all mRNA in muscle cytosol may be found in the protein-bound inactive form (Hershey, 1991). To respond rapidly, the mRNA can dissociate from the protein and be ready to act as a template for protein synthesis. Of course, the reverse could happen since active mRNA could be taken out of commission by protein binding. Because cellular control of mRNA translation can occur rapidly, cells can respond to an exercise training program or change in nutritional status by altering the amount of synthesis of specific proteins.

The initiation step is the major point of control of translation. Many initiation factors (eIFs) and several elongation factors (eEFs) can have their activity altered by the covalent attachment of phosphate groups. Addition of the phosphate group, donated by ATP, can either increase or decrease the ability of the initiation or elongation factors to participate in their specific steps in the translation process. Of course, detachment of phosphate groups would have the opposite effect. Phosphorylation/dephosphorylation of proteins also rapidly alters protein function in a cell in response to specific stimuli; we will discuss this process in a later section.

The lifetime of mRNA molecules, measured in half-life, can vary from 20 minutes to more than 24 hours and so must be considered as a level of control. As we have seen, mRNA is generated by transcription. How long the mRNA lasts in the cytosol is controlled by degradative enzymes known as ribonucleases. These enzymes cleave the sugar phosphate backbone of RNA, releasing individual nucleoside monophosphates. How long an individual mRNA molecule lasts in the cytosol depends on its being able to escape being degraded by the cytosolic ribonucleases. These ribonucleases may act at the 5′ end or the 3′ end of mRNA; some can even cleave in the middle. The presence of the 5′ cap protects the mRNA from the 5′ ribonucleases. A long poly A tail also seems to be a stabilizing signal because cleavage of mRNA from the 3′ end must first await shortening of the poly A tail down to 10 or so adenine nucleotides.

Posttranslational Processing

After the polypeptide is released from the translation apparatus, it is probably not in its physiologically active form or even in the cell location where it is to function. A group of proteins, known as molecular chaperones, bind to many completed polypeptide chains, help them assume their final three-dimensional conformation, help guide them to their site of action within the cytosol, or help them cross a membrane if they are to work in a cell organelle such as a mitochondrion.

Some polypeptides must also bind with other polypeptides to generate the active oligomeric protein. For example, myosin is a hexamer, so six polypeptides must combine in the proper way to generate a functional myosin molecule. Some polypeptides need to have a prosthetic group attached to them in order to be effective. For example, myoglobin, the protein in muscle that aids in storing and moving oxygen in the cytosol, needs to have a heme group attached to its polypeptide. Other polypeptides must be packaged for export because they function in the blood. Others must be incorporated into membranes because much of the mass of a membrane is made up of a variety of proteins. It is clear that the process of translation only generates a polypeptide component and that often many modifications must take place before that polypeptide can act in a concerted way in cellular function.

Summary

Protein synthesis occurs in the cell cytosol using mRNA, two ribosome subunits (described on the basis of their mobility in a centrifugal field as 40S and 60S), amino acids, energy sources, and a large number of protein factors. The modified rRNA molecules, created from two different kinds of RNA gene transcripts, are major components in the structure of ribosome subunits. A significant first step is the attachment of amino acids to specific tRNA molecules by a specific aminoacyl-tRNA synthetase. Each tRNA contains a three-base sequence called an anticodon; the complementary base pairing between the anticodon on tRNA (containing its attached amino acid) with the codon on mRNA provides fidelity in amino acid sequence in a protein.

Translation initiation is the most complex step in protein synthesis. It involves the formation of a complete ribosome subunit (80S) at the start (AUG) codon on mRNA with a methionyl-tRNA. Energy provided from GTP and the assistance of a number of eukaryotic initiation factors (eIFs) are also necessary. Elongation is the step where individual aminoacyl-tRNA molecules come to the ribosome complex and form a peptide bond between amino acids attached to their respective tRNAs. Individual ribosomes then move along the mRNA molecules three bases at a time. Each time the ribosome complex moves along the mRNA, a new amino acid is added to the growing polypeptide chain. Elongation continues the building of a polypeptide chain until a stop codon on mRNA is encountered. The ribosome complex then dissociates from the mRNA, and the completed polypeptide is released. Each mRNA molecule may have a number of ribosomes moving along independently, each with a growing polypeptide chain. Since the quantity of individual proteins in a cell can dictate rates of cell metabolism, translation, or protein synthesis, must be carefully regulated. Translation, mainly initiation, is controlled at a number of different sites. The lifetime of its mRNA molecules also determines the amount of a cell's protein.

Metabolism

Metabolism is the sum of all the chemical reactions that take place in a living organism. It is typically subdivided into reactions in which chemical substances are broken down to generate energy (known as catabolism) or reactions in which something larger is made out of smaller substances (known as anabolism). We have already seen anabolism in the synthesis of a polypeptide chain.

Anabolic reactions require an input of energy, which is provided when the energy released during catabolic reactions is harnessed in the formation of ATP. ATP is the currency used in living organisms, earned when fuel molecules are broken down and used to drive the anabolic part of metabolism.

This study of metabolism starts with a brief overview of the important elements of thermodynamics as they apply to living organisms. After this, we proceed to the basic but important chemical reactions involving carbohydrates, fats (known to biochemists as lipids), and amino acids.

CHAPTER 6

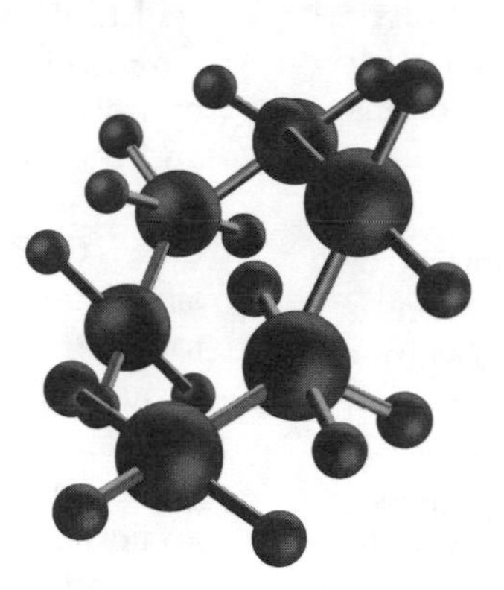

Biochemical Energetics

Metabolism, which represents all the chemical reactions or processes that occur in a living organism, operates in much the same way as a gasoline engine. For the engine to do useful work, it needs to oxidize fuel. For our bodies to grow and function (useful endeavors), we need to oxidize fuels. Our bodies and the gas engine are also open systems that exchange matter and energy with the surroundings. The engine takes in fuel and oxygen and gives off heat and exhaust gases and generates useful work. We take in food, water, and oxygen and give off heat, waste gases, liquids, and solids and, it is hoped, generate some useful work. We will first look at the energetic aspects of metabolism before focusing on the details of the various pathways and processes.

The Concept of Free Energy

Before we begin, however, let us review a few simple concepts:

- All chemical reactions involve energy changes.
- Chemical reactions in living organisms are catalyzed by enzymes.
- Enzymes attempt to drive the reaction they catalyze toward equilibrium.
- As enzyme-catalyzed reactions proceed toward equilibrium, they release energy.
- The farther a reaction is from equilibrium, the more energy it can release.
- Some energy released in a chemical reaction can be used to do useful work; the remainder is unavailable.

In thermodynamics, we focus only on differences between initial and final states; in this sense, we talk about changes. The following thermodynamic terms are defined as applied to biological processes:

Free energy change (ΔG): That part of the total energy change in a reaction or process that is capable of doing work at constant temperature and pressure. When free energy is released, it is considered by convention to be negative, and the reaction or process is said to be exergonic. When free energy must be added to a reaction or process, it is considered positive, and the reaction or process is said to be endergonic. Reactions or processes that release free energy (ΔG is negative) are considered to occur spontaneously, that is, by themselves. Reactions or processes that have a positive value for ΔG cannot occur by themselves and must be driven by a reaction or process that releases free energy.

Enthalpy change (ΔH): The total heat energy change in a reaction or process. When heat energy is given off or released in a reaction or process, we have a negative value for ΔH; such a reaction is said to be exothermic. Reactions or processes with a positive value for ΔH are said to be endothermic.

Entropy change (ΔS): Entropy is a measure of randomness or disorder. When a reaction or process has a positive value for ΔS, it becomes more disordered, that is, something is broken down. When a reaction or process has a negative value for ΔS, it becomes more organized. When something is made, as when a protein is synthesized from amino acids, ΔS is negative.

The relationship between enthalpy, entropy, and free energy for a reaction or process is given by the expression

$$\Delta G = \Delta H - T\Delta S.$$

(Equation 6.1)

This equation describes the relationship between the free energy change, the enthalpy change, and the entropy change for a reaction or process that occurs at the temperature T, given in degrees Kelvin (°K = °C + 273).

ATP hydrolysis, the oxidation of fuels, and glycolysis are examples of biological exergonic reactions, whereas protein synthesis, muscle contraction, creating ion gradients across membranes, and the storing of fuels are examples of endergonic processes.

As mentioned, endergonic reactions cannot occur by themselves but must be driven by a simultaneous exergonic reaction such that the combined endergonic plus exergonic reaction is exergonic. For example, a muscle cannot contract by itself; it must be driven by the simultaneous hydrolysis of ATP. Joining an amino acid to its proper tRNA is an endergonic reaction. However, if an ATP molecule is simultaneously hydrolyzed to AMP and PPi, aminoacyl tRNA forms, and the overall process is exergonic. Also, amino acids cannot be joined together to make a peptide unless GTP is simultaneously hydrolyzed; the overall process is exergonic. Glucose molecules cannot join together to make glycogen unless UTP is simultaneously hydrolyzed.

In summary, exergonic reactions can drive endergonic reactions provided the sum of the two is exergonic. In most of the examples given, the free energy released by the hydrolysis of nucleoside triphosphates (NTP), such as ATP, GTP, and UTP, drives endergonic reactions or processes. As shown in Figure 6.1, processes such as glycolysis or the oxidation of fuels are used to combine Pi and NDP (nucleoside diphosphate) to make NTP. Then the NTP drives the endergonic reaction. This process is the essence of living things.

Quantitative Values for Free Energy

In thermodynamics, one cannot measure exact values associated with an initial and final state of a reaction. However, we can measure energy changes over a reaction's course as changes in enthalpy, changes in free energy, and changes in entropy.

For the reversible chemical reaction given by the equation

$$A + B <\!\!\longrightarrow C + D,$$

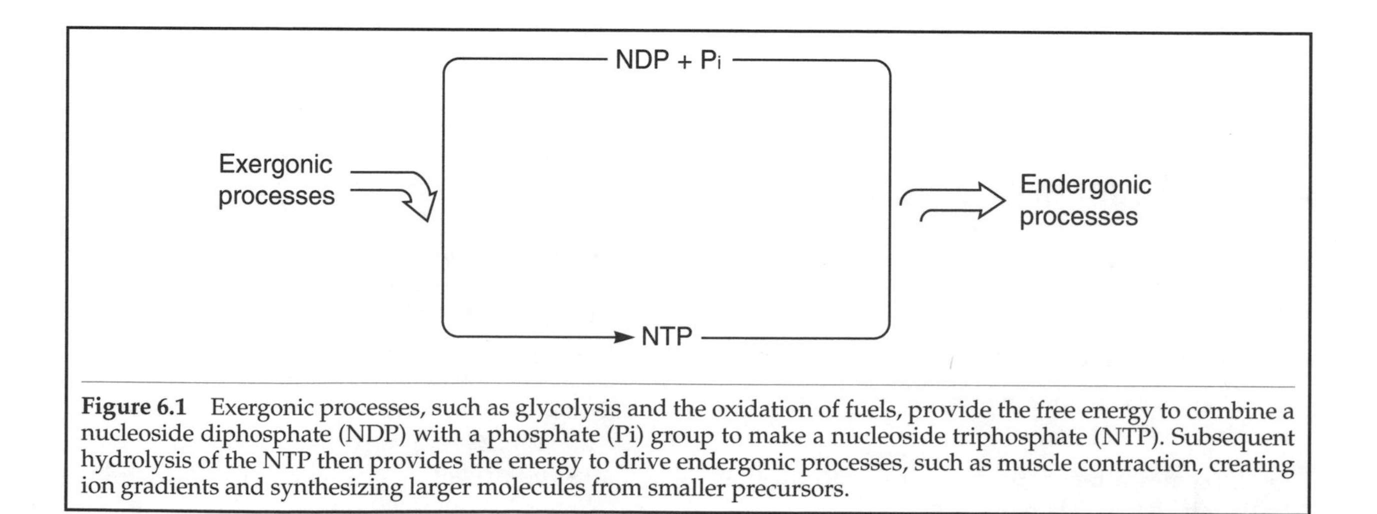

Figure 6.1 Exergonic processes, such as glycolysis and the oxidation of fuels, provide the free energy to combine a nucleoside diphosphate (NDP) with a phosphate (Pi) group to make a nucleoside triphosphate (NTP). Subsequent hydrolysis of the NTP then provides the energy to drive endergonic processes, such as muscle contraction, creating ion gradients and synthesizing larger molecules from smaller precursors.

the free energy change is given by the expression

$$\Delta G = \Delta G^{\circ\prime} + RT \ln \frac{[C]\,[D]}{[A]\,[B]}$$

(Equation 6.2)

where ΔG is the free energy change.

$\Delta G^{\circ\prime}$ is the standard free energy change for the reaction under what are considered to be standard conditions (when each of the reactants and products is at a concentration of 1.0 M). A prime is added to indicate a pH of 7.0; the traditional thermodynamic standard free energy is defined at a most unbiological pH of zero. $\Delta G^{\circ\prime}$ reflects the energy generating potential of the reaction, and a value for $\Delta G^{\circ\prime}$ can be determined from the equilibrium constant at a pH of 7.0 (K'_{eq}, discussed later).

R is the gas constant, which has a value of 8.314 joules per mole per °K.

T is the absolute temperature expressed in °K.

[A] is the concentration of species A, [B] is the concentration of species B, etc.

Suppose we start the above reaction by adding A and B together. The reaction will quickly proceed to the right, and as C and D accumulate, the net reaction (toward C and D) will gradually slow down. Eventually the forward (toward C and D) and backward reactions will occur at the same rate, and we will have reached equilibrium. From an energy perspective we know the following:

1. The farther a reaction is from equilibrium, the larger the value of ΔG (as a negative value). In the previous equation, when the reaction begins, [A] and [B] will be large, whereas [C] and [D] will be small. Accordingly, the natural logarithm (ln) term will be large and negative. Thus at the start of the reaction the ΔG will be large and negative.
2. As the reaction proceeds toward equilibrium, the numerical value for [A] and [B] will decrease, whereas the value for [C] and [D] will increase. Accordingly ΔG will decrease but still be negative.
3. At equilibrium, there is no net reaction and no net energy change, so the value for ΔG will be zero. Accordingly, a reaction at equilibrium releases no free energy.

Rewriting Equation 6.2 at equilibrium and $\Delta G = 0$, we get

$$0 = \Delta G^{\circ\prime} + RT \ln \frac{[C]\,[D]}{[A]\,[B].}$$

(Equation 6.2)

However, at equilibrium (and a pH of 7.0) the concentration ratio [C][D]/[A][B] will be given by the equilibrium constant K'_{eq}. We will use this fact and rearrange the equation:

$$\Delta G^{\circ\prime} = -RT \ln K'_{eq.}$$

(Equation 6.3)

This equation allows us to determine the value of the standard free energy change for a reaction at pH 7.0 if we know the temperature and the value of the equilibrium constant. We can say the following about $\Delta G^{\circ\prime}$ value for a reaction:

1. It is determined by the value for the equilibrium constant.
2. It reflects the energy-generating potential for the reaction.
3. It does not define the actual energy change for the reaction inside a cell.

The actual free energy change for a reaction inside a tissue is easy to determine. Take a sample of tissue and quickly freeze it, immediately stopping chemical reactions. Analyze chemically the amount of reactant(s) and product(s) for the reaction in question. From this, one can determine the mass action ratio for the reaction. This is the ratio of product to reactant concentrations as they exist at the moment the tissue was frozen, given by T—a Greek letter known as tau. The K'_{eq} for the reaction can be determined by allowing it to come to equilibrium in a test tube under conditions of temperature, pH, and ion concentration, identical to the tissue, and then analyzing the equilibrium concentrations of reactant(s) and product(s). Using Equation 6.4, we could determine the free energy change of our reaction as it existed at the time it was frozen.

$$\Delta G = -RT \ln \frac{K'_{eq}}{T}$$

(Equation 6.4)

Free energy changes for linked reactions are additive. Suppose we want to determine the standard free energy change for ATP hydrolysis.

$$ATP + H_2O \longrightarrow ADP + Pi$$

The reaction proceeds so far to the right that it is difficult to determine an equilibrium constant. Thus establishing an actual concentration for ATP at equilibrium would be impossible. However, we can determine this another way. Look at the two reactions below. In the first, ATP is used to phosphorylate (add a phosphate group to) glucose. In the second, glucose 6-phosphate, the product of this phosphorylation, is dephosphorylated. If we know the standard free energy changes for each of these reactions, and if we add these reactions together algebraically, the net reaction (under the line) is the reaction for ATP hydrolysis. The algebraic sum of the standard free energy changes is then the value for ATP hydrolysis.

1) ATP + glucose ——> glucose 6-phosphate + ADP
$\Delta G^{\circ\prime} = -23$ KJ/mole

2) glucose 6-phosphate + H_2O ——> glucose + Pi
$\Delta G^{\circ\prime} = -14$KJ/mole

---

net) ATP + H_2O ——> ADP + Pi
$\Delta G^{\circ\prime} = -37$ KJ/mole

Energy-Rich Phosphates

Living organisms obtain their energy from the breakdown of fuels such as fat, carbohydrate, and amino acids (from protein). Inside cells, fuel molecules are catabolized to simple products such as CO_2 and H_2O. During this catabolism, the free energy released drives the phosphorylation of nucleoside diphosphates to form nucleoside triphosphates. The major nucleoside diphosphate is ADP, and the product is ATP. The reaction illustrating the formation of ATP is

ADP + Pi ——> ATP + H_2O.

The hydrolysis of ATP (or GTP or UTP) drives endergonic reactions or processes. Notice that the reactions showing the formation of ATP and the hydrolysis of ATP are exactly reversed, although these are definitely not reversible reactions. Each day, an active person turns over (breaks down and resynthesizes) an amount of ATP almost equivalent to double body weight.

The structure of ATP, shown in shorthand in Figure 6.2, contains three phosphate groups, joined by two anhydride bonds. Although the phosphate groups are stabilized by magnesium in the cell, negative charges are still in close proximity and breaking the bonds between either the α and β or β and γ phosphate groups is energetically favorable. Thus the hydrolysis of anhydride bonds of ATP generates energy to drive endergonic reactions. Extra energy is obtained when the hydrolysis is between the α and β phosphates, because the product PPi can also be hydrolyzed by an ubiquitous enzyme, inorganic pyrophosphatase, which results in a further large free energy release (see Figure 6.2).

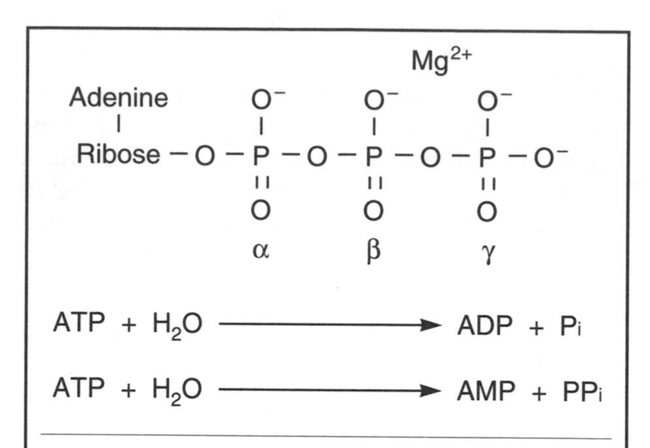

Figure 6.2 A simplified structure for ATP emphasizing the three phosphate groups often identified with Greek letters. In the cell, ATP is associated with the positively charged magnesium ion (Mg^{2+}). Hydrolysis of ATP between the α and β phosphates (top equation) or between the β and γ phosphates (second equation) generate free energy release in the cell. Inorganic pyrophosphate (PPi) can also be hydrolyzed to inorganic phosphate (Pi) by a ubiquitous enzyme, inorganic pyrophosphatase.

ATP would be useless as an energy currency in the cell if the concentrations of ATP, ADP, and Pi were equilibrium concentrations. At equilibrium, the concentration of ADP would be more than 10 million times greater than ATP, with no net ATP hydrolysis and no free energy release. For ATP to be the important energy currency it is, the concentration in the cell is kept very far from equilibrium so that the ratio [ATP]/[ADP] is normally greater than 50. Moreover, when ATP is used as an energy source it is replenished at the same rate so that its concentration does not decrease. Because large free energy changes accompany the hydrolysis of ATP and the other nucleoside triphosphates, they are known as energy-rich phosphates.

The Pool of Phosphates in the Cell

The adenine nucleotides (i.e., ATP, ADP, AMP) are primarily involved in coupling exergonic and endergonic reactions. ATP is formed from ADP when fuel molecules are broken down, and ATP hydrolysis drives most endergonic processes. GTP drives peptide bond formation during protein synthesis, and one GTP molecule is formed during the citric acid cycle. Also, UTP is used to make glycogen in liver and muscle. We must also remember that GTP, UTP, CTP, and ATP are used to make the various molecules of RNA. Furthermore, when RNA molecules are broken down, the products are NMP molecules (GMP, UMP, CMP, and AMP).

This means that there must be specific reactions to interconvert the various components of the nucleoside phosphate pool. For example, nucleoside diphosphates must convert to nucleoside triphosphates and nucleoside monophosphates to nucleoside diphosphates. For the first interconversion, a nonspecific enzyme known as nucleoside diphosphate kinase transfers phosphate groups to and from ATP/ADP and NTP/NDP, as follows:

$$\text{NDP} + \text{ATP} \xleftrightarrow[Mg^{2+}]{\text{nucleoside diphosphate kinase}} \text{NTP} + \text{ADP}$$

In this freely reversible reaction, NDP (such as UDP, GDP, and CDP) is phosphorylated to NTP by accepting a phosphate group from ATP. The magnesium ion under the arrow is always present in reactions involving nucleoside triphosphates (see Figure 6.2). Because the reaction is freely reversible, its net direction depends on the relative concentrations of ATP, ADP, and the other NTPs and NDPs. The overall effect of the nucleoside diphosphate kinase reaction is to maintain a balance in the ratio of NTP/NDP in a cell.

To convert NMP to NDP, nucleotide-specific enzymes transfer a phosphate group from ATP to the nucleoside monophosphate, making ADP and a nucleoside diphosphate. They are said to be specific because one exists for each nucleoside monophosphate. For example, AMP kinase catalyzes the following reaction:

$$\text{AMP} + \text{ATP} \xleftrightarrow[Mg^{2+}]{\text{AMP kinase}} \text{ADP} + \text{ADP}$$

UMP kinase catalyzes a similar reaction:

$$\text{UMP} + \text{ATP} \xleftrightarrow[Mg^{2+}]{\text{UMP kinase}} \text{ADP} + \text{UDP}$$

These freely reversible reactions also need magnesium ions because these ions are bound to the ATP molecules. The net effect of the nonspecific nucleoside diphosphate kinase and the specific nucleoside monophosphate kinase (e.g., AMP kinase and UMP kinase) is to maintain a balance between the NTP/NDP/NMP in the cell.

Phosphagens

The ATP concentration in most tissues is fairly low, about 3 to 8 mM. Since ATP represents the immediate energy source to drive endergonic processes, problems could arise if ATP is rapidly used up. In cells with a slow acceleration of ATP-consuming reactions, ATP concentration can be easily maintained by a gradual acceleration of ATP-producing reactions, such as fuel oxidation. However, in muscle, this could be a big problem because, during sprinting, ATP can be hydrolyzed at the rate of about 4 mmoles per kg of muscle per second. Thus, all the muscle cells' ATP could be consumed in about 2 seconds, if not regenerated.

ATP in cells is regenerated from ADP by breaking down fuel molecules using aerobic or anaerobic catabolic processes. However, ATP-regenerating processes cannot produce ATP at the same rate at which it is hydrolyzed to drive muscle contraction during sprinting. Moreover, these processes take time to gear up to maximum speed, whereas at the start of a sprint, the rate ATP is hydrolyzed is about maximal. To prevent muscle cells from using up their ATP at the start of maximal or near maximal contractions, an alternate energy-rich molecule, known as a phosphagen, is capable of regenerating ATP at a very high rate. In vertebrate muscle the phosphagen is creatine phosphate (abbreviated CP), also called phosphocreatine (abbreviated PCr). In some invertebrate muscles, the phosphagen is arginine phosphate.

Creatine phosphate has its phosphate group transferred to ADP to yield ATP and creatine, catalyzed by an enzyme known as creatine kinase:

$$\text{ADP} + \text{CP} \xleftrightarrow[Mg^{2+}]{\text{creatine kinase}} \text{ATP} + \text{C}$$

This reaction is freely reversible. During muscle contraction, the forward direction is favored in order to regenerate ATP. During recovery, the backward reaction is favored to regenerate CP. The activity of creatine kinase is very high in muscle in order to match the regeneration of ATP during the most vigorous muscle contractions. The actual concentration of CP in muscle is about 4 to 5 times that of ATP (about 18 mmoles per kg of muscle), not that much considering how fast ATP can be used in a vigorously contracting muscle. However, it is enough to act as a temporary ATP buffer until other ATP-regenerating processes reach maximal rates. The ATP level in muscle must not be allowed to drop very much, for if it does, a condition known as rigor occurs, which can be damaging to muscle cells. In fact, rigor mortis is caused by the loss of muscle ATP some time after death, as a result of the inability of the muscle cells to regenerate ATP.

Muscle Contraction

Skeletal muscle, the largest tissue in the body, consists of a number of elongated cells known as fibers. Neurons innervate each fiber, and nerve impulses crossing from the neuron to the muscle fiber membrane (or sarcolemma) activate the fiber to make it contract. The steps in this process may be summarized as follows:

- The neuron signal spreads a wave of depolarization over the sarcolemma and down into the interior of the fiber via surface invaginations known as T-tubules.
- T-tubule depolarization is linked to calcium ion (Ca^{2+}) release from a specialized form of endoplasmic reticulum known as sarcoplasmic reticulum or SR.
- The resulting increase in calcium ion concentration in the fiber interior is roughly 10^{-7} M to about 10^{-5} M (a 100-fold increase).
- The calcium ions bind to a protein (troponin) in a thin filament in the fiber.
- This binding allows protein projections on another filament, known as the thick filament, to bind to the thin filament.
- When these projections or cross bridges bind to the thin filament, they pull on it and cause the muscle fiber to shorten.
- The action of millions of cross bridges in many fibers causes the overall muscle to shorten and do work.
- The energy to make individual cross bridges attach to actin and cause shortening is derived from the hydrolysis of ATP.

The thick filament is composed predominantly of a protein known as myosin (see Figure 1.15). The cross bridges are part of the myosin molecule. The protein in the thin filament that gets attached by the myosin cross bridges is known as actin. Myosin has two major properties: (a) the ability to bind actin, and (b) the ability to hydrolyze ATP as an enzyme. The ATP-hydrolyzing activity of myosin is known as ATPase activity. Actin also has two important properties: (a) myosin-binding and (b) an ability to activate the ATPase activity of myosin. When a myosin cross bridge binds to actin its ATPase activity is increased. We call this actin-activated myosin ATPase activity or actomyosin ATPase activity, illustrated as follows:

$$\text{ATP} + H_2O \xrightarrow[\text{Mg}^{2+}]{\text{actomyosin ATPase}} \text{ADP} + \text{Pi}$$

ATPases hydrolyze ATP so that work can be done. In muscle contraction, ATP hydrolysis by the actomyosin ATPase allows the muscle to shorten. Two other ATPases are also involved in muscle contraction. One is the sodium-potassium ATPase found in muscle sarcolemma. Its function is to pump sodium ions out of the cell and potassium ions back in after the muscle membrane depolarization ends, using energy derived from the hydrolysis of the ATP. The other ATPase, in the SR, hydrolyzes ATP and uses the free energy released to pump calcium ions back into the SR when the muscle is to relax; it is known as the SR Ca^{2+} ATPase.

The contraction of muscle involves ATP hydrolysis at three locations. Of course, this ATP must be regenerated quickly because, as previously stated, little ATP is stored in muscle. The ATP concentration in muscle is effectively maintained by three major processes, illustrated in Figure 6.3. The ATPases represent the actomyosin ATPase; the ATPase associated with the sarcolemma to pump sodium and potassium ions and the ATPase that pumps calcium ions back into the SR. ATP can be regenerated by the creatine kinase reaction, the oxidation of fuels (oxidative phosphorylation), and glycolysis.

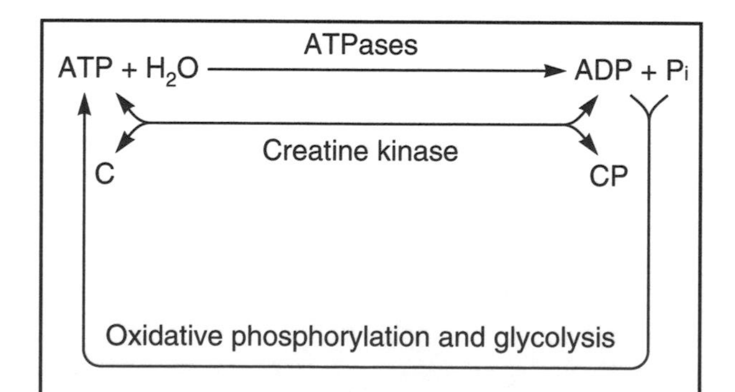

Figure 6.3 In muscle, three ATP-hydrolyzing enzymes (ATPases) use ATP at a rapid rate. The ATP concentration diminishes only slightly due to the rapid response of the creatine kinase reaction and the aerobic and anaerobic (glycolysis) breakdown of fuel molecules. Although the creatine kinase reaction is freely reversible, during muscle contraction, the net direction at the site of the ATPase enzymes is to regenerate ATP at the expense of creatine phosphate (CP).

The creatine kinase reaction can regenerate ATP rapidly because the activity of the enzyme is so high. Creatine kinase thus has a high power for regenerating ATP. However, because of a limited supply of CP, this process has a low capacity.

Glycolysis is a sequence of 11 enzyme-catalyzed reactions in which glucose or glycogen stored in the muscle is converted to lactate. At the same time, ATP is regenerated from ADP. The reaction that follows outlines the stoichiometry of glycolysis when glucose is the starting fuel:

$$\text{glucose} + 2\,\text{ADP} + 2\,\text{Pi} \longrightarrow 2\,\text{lactate} + 2\,\text{H}^+ + 2\,\text{ATP}$$

In this process, one glucose molecule is converted to two lactate ions and two hydrogen ions, and the free energy released is used to phosphorylate two ADP with two Pi to make two ATP. This process has a moderate power and capacity to generate ATP, and with no direct involvement of molecular oxygen, the process is anaerobic.

Oxidative phosphorylation is the formation of ATP from ADP and Pi in association with the transfer of electrons from fuel molecules to coenzymes to oxygen. The products of oxidative phosphorylation are H_2O and CO_2. This process has a low power but a high capacity because a major fuel for oxidative phosphorylation is fat. Even a very lean person has enough stored fat to provide fuel for many hours of oxidative phosphorylation. Figure 6.4 outlines the overall scheme of oxidative phosphorylation. In this scheme, SH_2 represents fuel molecules. Electrons associated with the hydrogen are transferred from SH_2 to coenzymes (represented by NAD^+) and then the electrons on the now reduced coenzyme (NADH) are transferred on to oxygen, forming H_2O. During this process, enough free energy is generated to phosphorylate ADP with Pi to make ATP.

Before leaving the topic of muscle, two other reactions must be considered. The first is one we have seen previously. It is written here in the direction opposite to the way it was shown earlier, but this does not matter because the reaction is freely reversible with an equilibrium constant near one.

$$2\,\text{ADP} \underset{\text{Mg}^{2+}}{\overset{\text{adenylate kinase}}{\rightleftharpoons}} \text{ATP} + \text{AMP}$$

The enzyme for this reaction is named AMP kinase, or in muscle, adenylate kinase or myokinase. You should be familiar with all three names.

The significance of this reaction is as follows: During hard muscle activity, the rate of ATP hydrolysis is high. Although muscle efficiently regenerates ATP from ADP by the three processes previously described, nonetheless an increase in ADP concentration exists when compared to a rested muscle. What the adenylate kinase (or AMP kinase or myokinase) does is cause two ADP molecules to interact and make one AMP and one ATP. This interaction keeps the ADP concentration from building up.

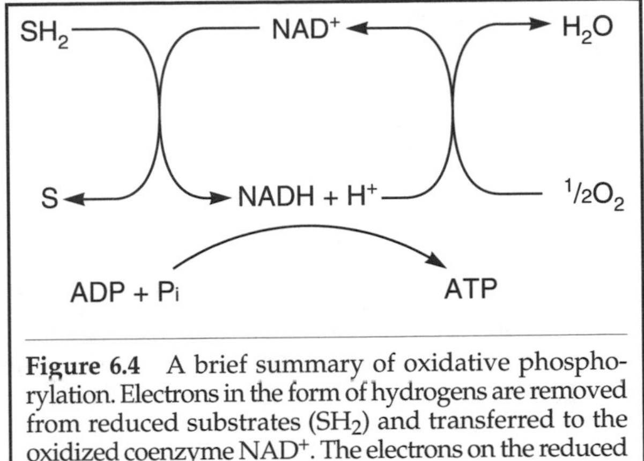

Figure 6.4 A brief summary of oxidative phosphorylation. Electrons in the form of hydrogens are removed from reduced substrates (SH_2) and transferred to the oxidized coenzyme NAD^+. The electrons on the reduced coenzyme NADH are transferred through a series of carriers (not shown) to oxygen, forming water. During the process of electron transfer, free energy is released to phosphorylate ADP, making ATP.

In the second reaction, an enzyme in skeletal muscle (especially the fast twitch or type II fibers) deaminates (removes the amino group) from the adenine of AMP. The amino group comes off as ammonia, NH_3, which would become protonated as the ammonium ion (NH_4^+), as shown:

$$\text{AMP} + \text{H}_2\text{O} \xrightarrow{\text{adenylate deaminase}} \text{IMP} + \text{NH}_3$$

Adenylate deaminase catalyzes the irreversible deamination of AMP, converting it into IMP (inosine monophosphate). This reaction works in concert with the adenylate kinase reaction during vigorous muscle work. In effect, two ADP molecules are converted into an ATP and an AMP by the adenylate kinase reaction. Then in the adenylate deaminase reaction, the AMP is irreversibly converted into IMP. These two related reactions have two major functions. First, the irreversible adenylate deaminase reaction drives the reversible adenylate kinase reaction to the right, to decrease [ADP], or, put another way, to keep [ADP] from building up. This reaction also maintains a high [ATP]/[ADP] ratio, which is important for the hydrolysis of ATP, for it results in a high free energy. If ADP concentration were too high, less free energy would be available to drive endergonic reactions, and if the free energy change was too little, it would be ineffective in driving endergonic processes. Furthermore, if ADP concentration increased too much, it could slow down ATPase reactions by a process known as product inhibition. Second, the product ammonia in the adenylate deaminase reaction is a base (i.e., a proton acceptor) and gets protonated to the ammonium ion NH_4^+. Formation of ammonium ion from ammonia removes one proton; this protonation keeps muscle from becoming too acidic (i.e., low pH). Moreover, the ammonium ion stimulates the process of glycolysis by activating one of the rate-controlling enzymes.

The adenylate deaminase reaction is only important when muscles are working hard. If you walked 800 m, this would not result in a significant reaction. However, if you ran 800 m as fast as possible, this reaction (in association with adenylate kinase) would be significant. Fast running uses the FT fibers to a major extent. The combined effect of these two reactions (adenylate kinase and adenylate deaminase) would allow you to get more out of the muscles as they gradually fatigued.

Suppose we took a muscle sample before and after running 800 m as fast as possible. If we made chemical measurements on the before and after muscle samples, we would note that, compared to the rested muscle sample, the post-800 m sample (i.e., fatigued muscle) would have the following differences:

- A high muscle lactate concentration, related to the fact that glycolysis is an important route for generating the needed ATP.
- A much lower pH. Glycolysis produces one proton for each lactate formed. Because of the protonation of ammonia, the pH decrease would be less than if the adenylate deaminase reaction were not taking place. Low muscle pH is an important reason why a fatigued muscle is forced to stop working.
- A greatly reduced concentration of CP. During the hard sprint, CP would be an important source for regenerating ATP.
- A decrease in TAN (total adenine nucleotides—i.e, the sum of [ATP] + [ADP] + [AMP]) and an increase in [IMP] and [NH_4^+]. The decrease in TAN and the increase in IMP and ammonium ions results from the combined effect of the adenylate kinase and adenylate deaminase reactions.
- A decrease in muscle glycogen concentration. Glycogen is an important fuel for glycolysis and thus would be used during the 800 m run.
- An increase in total Pi concentration, and in addition, an increase in the ratio [$H_2PO_4^-$]/[HPO_4^{2-}]. Since the CP concentration would decline (leading to an increase in creatine), there would be a corresponding increase in Pi. In addition, because the muscle pH would decline, the dihydrogen form of Pi would increase due to protonation of the more basic, monohydrogen form.

Summary

The thousands of chemical reactions in our bodies involve energy changes. Some release energy while others require an input of energy. The part of the total energy change from chemical reactions that can be harnessed to do useful work is known as free energy (ΔG). In spontaneous or exergonic reactions, free energy is given off, that is, ΔG is negative. Reactions in the body that require free energy (endergonic reactions) are driven by exergonic reactions. Molecules known as energy-rich phosphates, such as ATP and GTP, are used to drive endergonic reactions and processes because when one or more of the phosphate groups is removed by hydrolysis, much free energy is released. For most endergonic processes, ATP is the energy currency.

In muscle, a tissue that varies widely in its need for energy, the ATP concentration is only sufficient to drive about two seconds of maximal work. Therefore rapid ways to replenish the ATP must exist. Transfer of a phosphate group from creatine phosphate (CP) is a mechanism to rapidly renew ATP, but supplies of CP are limited. Glycolysis, an anaerobic process in which glycogen or glucose is broken down to lactate, can provide ATP at a fairly rapid rate, but there is a limited capacity for this process. Oxidative phosphorylation, however, provides ATP at a low rate for a prolonged period of time. Exercise at a high rate to fatigue can lead to severe metabolic displacement in muscle as the contracting fibers attempt to generate ATP to match its rate of breakdown by the ATPases. Severe decreases in CP and pH and large increases in lactate, ammonium, and phosphate ions characterize the fatigued state.

Chapter 7

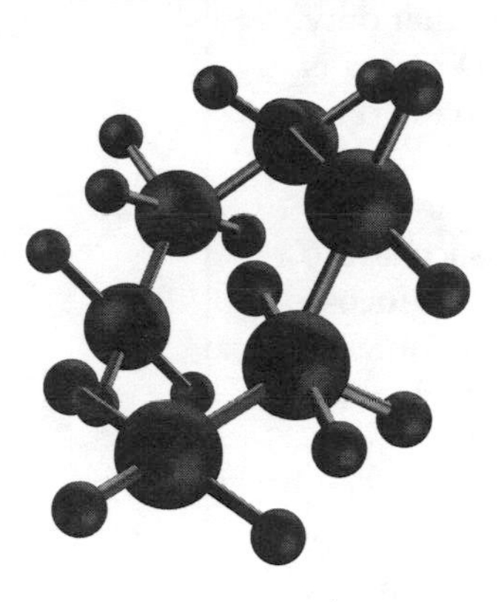

Glycolysis

Because the chapter on glycolysis represents our first detailed study of a metabolic pathway, we will begin with a general overview of the metabolism of the cells.

Overview of Metabolism

We have already described briefly the two major energy-generating pathways important to muscle: glycolysis and oxidative phosphorylation. The latter is the most important pathway for ATP regeneration, supplying about 85% of our ATP. Oxidative phosphorylation can utilize carbohydrate, fat, and amino acids, whereas glycolysis is restricted to carbohydrate as the fuel. The amino acids that are oxidized are obtained from the normal breakdown of proteins or from excess proteins in the diet that are not immediately converted into fat or glucose.

Most fat is stored in specialized cells known as fat cells; small amounts are also stored in other cells, such as muscle. When metabolism needs to be increased, nerve and hormonal signals cause fat to be broken down, and fatty acids are released to the blood for cells to use as fuel.

The body carbohydrate stores are not large. Some is stored in liver as glycogen, which is really a polymer made of individual glucose molecules joined together. Glycogen is also stored in muscle. The other form of carbohydrate in the body is the glucose in blood and extracellular fluids. Glucose is normally the sole fuel for the brain, the tissue that the body treats as most important. Glucose is also the only fuel used by red blood cells and part of the kidney and is needed by other tissues. It is the precursor needed to store fatty acids in fat cells. Because a need for glucose always exists, and not much is available (only as stored glycogen and glucose in extracellular fluids), it should not be surprising that its use is carefully regulated.

Most of the body's glucose comes from the ingestion of carbohydrates in the form of starch, sucrose, lactose, and free glucose and fructose in such foods as fruit. The body also has a limited capacity to make glucose from noncarbohydrate sources, such as amino acids, lactate, and glycerol. Gluconeogenesis is the process of making glucose from noncarbohydrate sources. This process is important when the body needs carbohydrate, but normal supplies in the body are low.

We classify carbohydrates as monosaccharides, disaccharides, and polysaccharides. Examples of monosaccharides are glucose, fructose, and galactose, simple sugars containing six carbon atoms called hexoses—*hex* for six carbon atoms and *ose* meaning sugar. Figure 7.1 shows the structures for the predominant forms of D-glucose, D-galactose, and D-fructose. The D refers to the configuration about carbon atom five in each molecule. Only the D forms of the monosaccharides are acceptable to glucose-metabolizing enzymes in animals. Recall that only L amino acids can be used by animals. D-ribose is also a monosaccharide but contains only five carbon atoms and would be known as a pentose.

Disaccharides are formed when two monosaccharides join together. The common disaccharide sucrose is composed of the monosaccharides glucose and fructose, while lactose or milk sugar contains glucose and galactose. When disaccharides are digested, the products are monosaccharides. Glycogen and starch are polysaccharides, but only starch is significant as a dietary source of carbohydrate. From a nutritional perspective, starch is a complex carbohydrate. When it is completely digested, its products are glucose molecules.

After a meal containing a variety of foods, the main carbohydrate digestion products would be glucose, some fructose, and galactose from milk sugar. These substances are absorbed into the blood and transported to the liver, where galactose is converted to glucose. Fructose can be converted into a phosphorylated form of glucose known as glucose 6-phosphate.

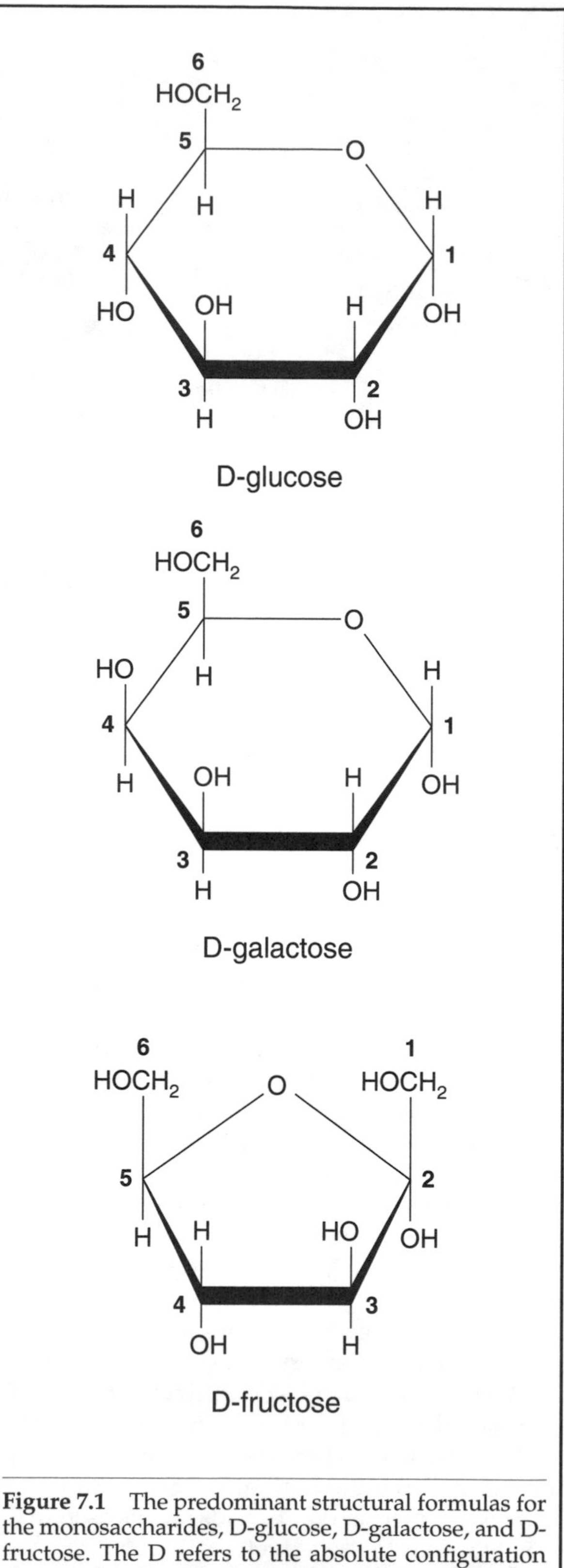

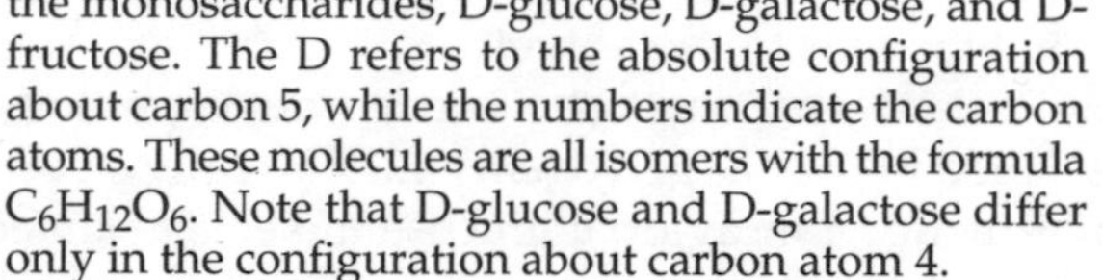

Figure 7.1 The predominant structural formulas for the monosaccharides, D-glucose, D-galactose, and D-fructose. The D refers to the absolute configuration about carbon 5, while the numbers indicate the carbon atoms. These molecules are all isomers with the formula $C_6H_{12}O_6$. Note that D-glucose and D-galactose differ only in the configuration about carbon atom 4.

Cellular Uptake of Glucose

The normal blood glucose concentration, described as euglycemia, is approximately 5 mM, equivalent to 90 mg of glucose per dL (100 mL) of blood. Following a meal, blood glucose is elevated above normal and can increase to 9 mM. Uncontrolled insulin-dependent diabetes mellitus or noninsulin-dependent diabetes mellitus can result in a glucose concentration over 20 mM. We call elevated blood glucose concentrations hyperglycemia. Blood glucose concentration well below normal (about 2.5 mM or less) is called hypoglycemia. You can become hypoglycemic if you do not eat for a long period of time, or if you exercise for hours without ingesting carbohydrate.

A gradient exists for glucose entry into cells because the glucose concentration in the blood and extracellular fluid is higher than inside cells. Glucose is a polar molecule, with 5 hydroxy (OH) groups and thus hydrophilic. It is therefore a poor substance for crossing the hydrophobic cell membrane. To get glucose inside cells, a transporter is needed; that is, a protein molecule that will allow glucose entry across the cell membrane. The process of transporting a substance down its concentration gradient across a membrane is known as facilitated diffusion. Membrane transport will be characterized with a

V_{max} and a K_m and will exhibit saturation kinetics similar to the effect of increasing substrate concentration on the activity of an enzyme. The five different kinds of glucose transporters, GLUT1, GLUT2, GLUT3, GLUT4, and GLUT5, all differ in their kinetic parameters and are found in specific tissues.

Glucose must enter a variety of cells in order to be used. Entry into some cells is regulated; entry into others is unregulated. We would expect the unregulated entry of glucose, which depends on the relative concentration gradient of glucose across the membrane, to occur in cells that rely primarily on glucose as an energy source. In fact, this is what happens for red blood cells, brain cells, and kidney cells. Liver cells, which store excess glucose as glycogen, also have unregulated glucose uptake. Large tissues, such as skeletal muscle or fat as well as the heart, have regulated glucose uptake. Glucose transport across cell membranes in regulated tissues occurs primarily by way of GLUT4, which, unlike the other transporter isoforms, is regulated by insulin.

Insulin, a polypeptide hormone secreted by the β cells of the pancreas, is the main regulator of glucose transport. When blood glucose concentration is elevated (e.g., following a meal), blood insulin concentration increases to help glucose enter the regulated tissues. Insulin binds to an insulin receptor, resulting in the attachment of a phosphate group from ATP onto a number of protein molecules. Phosphorylation of the proteins alters their properties, giving rise to the diverse effects of insulin in tissues. One main effect is to induce translocation of GLUT4 transporters from intracellular storage sites to the cell membrane to aid glucose entry. Thus, insulin increases the V_{max} of glucose transport but only in those cell types (i.e., muscle and fat) expressing the GLUT4 transporter gene. Exercising skeletal muscle also has an increased ability to take up glucose from the blood, independent of the effect of insulin. This muscle contraction effect persists into the early postexercise period in order to rebuild depleted stores. During prolonged exercise tasks or games, one must ingest glucose to keep blood levels maintained because exercising muscle has an augmented capacity to take up glucose from the blood. Failure to supply enough glucose to the body during prolonged physical activity can lead to problems associated with hypoglycemia.

Phosphorylation of Glucose

Once glucose enters a cell, it is covalently modified by transfer of the terminal phosphate from ATP to carbon atom six of glucose to make glucose 6-phosphate (glucose 6-P) as shown in the equation that follows. When glucose is phosphorylated, the product, glucose 6-P, is trapped inside the cell.

$$\text{glucose} + \text{ATP} \xrightarrow[\text{Mg}^{2+}]{\text{hexokinase/ glucokinase}} \text{glucose 6-P} + \text{ADP}$$

This reaction is essentially irreversible because of the large free energy change, as previously shown. Four isoenzymes catalyze this reaction, identified as hexokinases I, II, III, and IV (HK I, HK II, HK III, and HK IV). Hexokinase IV is also known by the common name glucokinase and is found in the liver.

The differences between glucokinase (known both by GK and HK IV) and the other hexokinase isozymes (HK I, HK II, and HK III) are as follows:

1. Glucokinase is found only in the liver and pancreas, whereas the other hexokinase isozymes are found in all cells.
2. The amount of hexokinase I, II, and III in cells remains fairly constant; they are thus constitutive enzymes. The amount of glucokinase in liver cells depends on the amount of glucose in the large portal vein that leads from the intestines to the liver or on the level of blood insulin. Glucokinase is thus described as an inducible enzyme because a diet high in carbohydrate will induce liver cells to make more. Conversely, diabetes (with its attendant low insulin), starvation, or a low carbohydrate diet will mean less glucokinase. Control of the amount of liver glucokinase is mainly at the transcription level since a 30- to 60-fold increase in gene transcription follows within 30 minutes after injection of insulin into a diabetic rat.
3. Hexokinase isozymes I, II, and III have a low K_m for glucose (0.02-0.1 mM), whereas glucokinase has a high K_m for glucose (about 5 mM or more). Hexokinase is thus very sensitive to glucose, whereas glucokinase activity only becomes important for phosphorylating glucose when the concentration of glucose is high inside liver cells.
4. Hexokinase isozymes I, II, and III, but not glucokinase, can be inhibited by the product of the reaction, glucose 6-P. Thus, if the concentration of glucose 6-P increases inside a cell, it creates feedback which inhibits the activity of hexokinase, glucose will not get phosphorylated, and its concentration will increase in the cell. This increase will reduce

the gradient for transport into the cell, thus slowing down glucose entry.

5. Hexokinase I, II, and III can phosphorylate other hexoses, whereas glucokinase can only phosphorylate glucose.

Reactions of Glycolysis

Strictly speaking, glycolysis means the splitting of glucose, summarized as follows:

$$\text{glucose} + 2\ \text{ADP} + 2\ \text{Pi} \longrightarrow 2\ \text{lactate} + 2\ \text{H}^+ + 2\ \text{ATP}$$

This reaction occurs in the cytoplasm of cells. Tissues such as liver, skeletal muscle, and heart muscle contain glycogen. When glycogen is initially broken down, it produces glucose 6-P, which can be fed into the reactions of glycolysis. Figure 7.2 shows an overview of glycolysis.

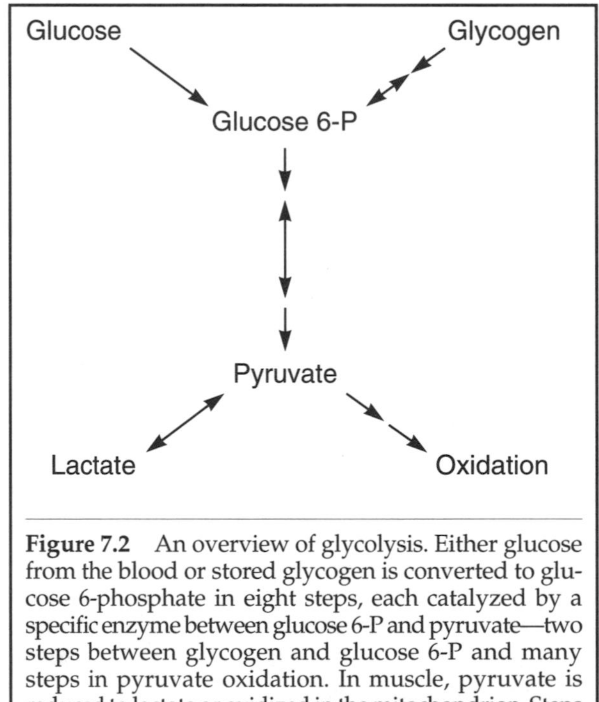

Figure 7.2 An overview of glycolysis. Either glucose from the blood or stored glycogen is converted to glucose 6-phosphate in eight steps, each catalyzed by a specific enzyme between glucose 6-P and pyruvate—two steps between glycogen and glucose 6-P and many steps in pyruvate oxidation. In muscle, pyruvate is reduced to lactate or oxidized in the mitochondrion. Steps shown with a double arrow are considered reversible.

Either glucose or glycogen can give rise to glucose 6-P. This process is then followed by a series of eight reactions, finally producing pyruvate. The pyruvate can have two major fates: (a) to be reduced to lactate or (b) to enter the mitochondria for oxidation to carbon dioxide and water. Because glycolysis is considered to be an anaerobic process, strictly speaking glycolysis must end with lactate, since pyruvate oxidation absolutely needs oxygen. To avoid this semantic problem, the term aerobic glycolysis is used for the process in which the pyruvate is oxidized in the mitochondrion. We will look at the mitochondrial reactions in a later chapter.

Controversy exists about whether the enzymes of glycolysis are freely dissolved in the cytosol of the cell. Many, if not all, may be bound to other structures in the cell, such as structural filaments providing shape to the cell, to the endoplasmic reticulum (sarcoplasmic reticulum in muscle), to the outer membrane of the mitochondrion (as is the case for hexokinase), or to contractile proteins such as actin. Some of the enzymes catalyzing the reactions of glycolysis may also be physically linked to each other such that the product of one enzyme is immediately passed to the next enzyme as its substrate. In this case, the actual concentrations of the intermediates in glycolysis, except glucose 6-P, fructose 6-P, pyruvate, and lactate, would not increase much, even if glycolysis were proceeding rapidly.

Glycolysis has two major functions. One is to generate energy in the form of ATP. In fact, in red blood cells, in which glycolysis is the only energy-generating source, making ATP is its only role. The second function is to generate pyruvate for final oxidation in the mitochondrion. Figure 7.3 outlines the major reactions of the glycolytic pathway.

Discussion of the individual reactions of glycolysis will start with glucose 6-P, since glucose phosphorylation has already been described. The glucose phosphate isomerase reaction interconverts two hexose phosphates. This is called an isomerase because glucose 6-P and fructose 6-P are isomers.

The next step is the phosphorylation of fructose 6-P, using a phosphate group from ATP and catalyzed by the enzyme phosphofructokinase (abbreviated PFK). The phosphate group is attached to carbon 1 of fructose, so the product is known as fructose 1,6-bisphosphate. The term bis means there are two phosphates attached to separate locations on the same molecule. If there were three phosphates attached to separate places on the same molecule, we would use the term tris. PFK catalyzes the committed step of glycolysis, committing the cell to glucose degradation. PFK is under tight regulation, and its activity controls the flux of glycolysis, or the overall rate at which it occurs.

Aldolase splits a hexose bisphosphate (fructose 1,6-bisphosphate) into two triose phosphates (three-carbon sugars, each one with a phosphate group; i.e., glyceraldehyde 3-phosphate and dihydroxyacetone phosphate). However, only glyceraldehyde 3-phosphate has a further role in glycolysis. Therefore,

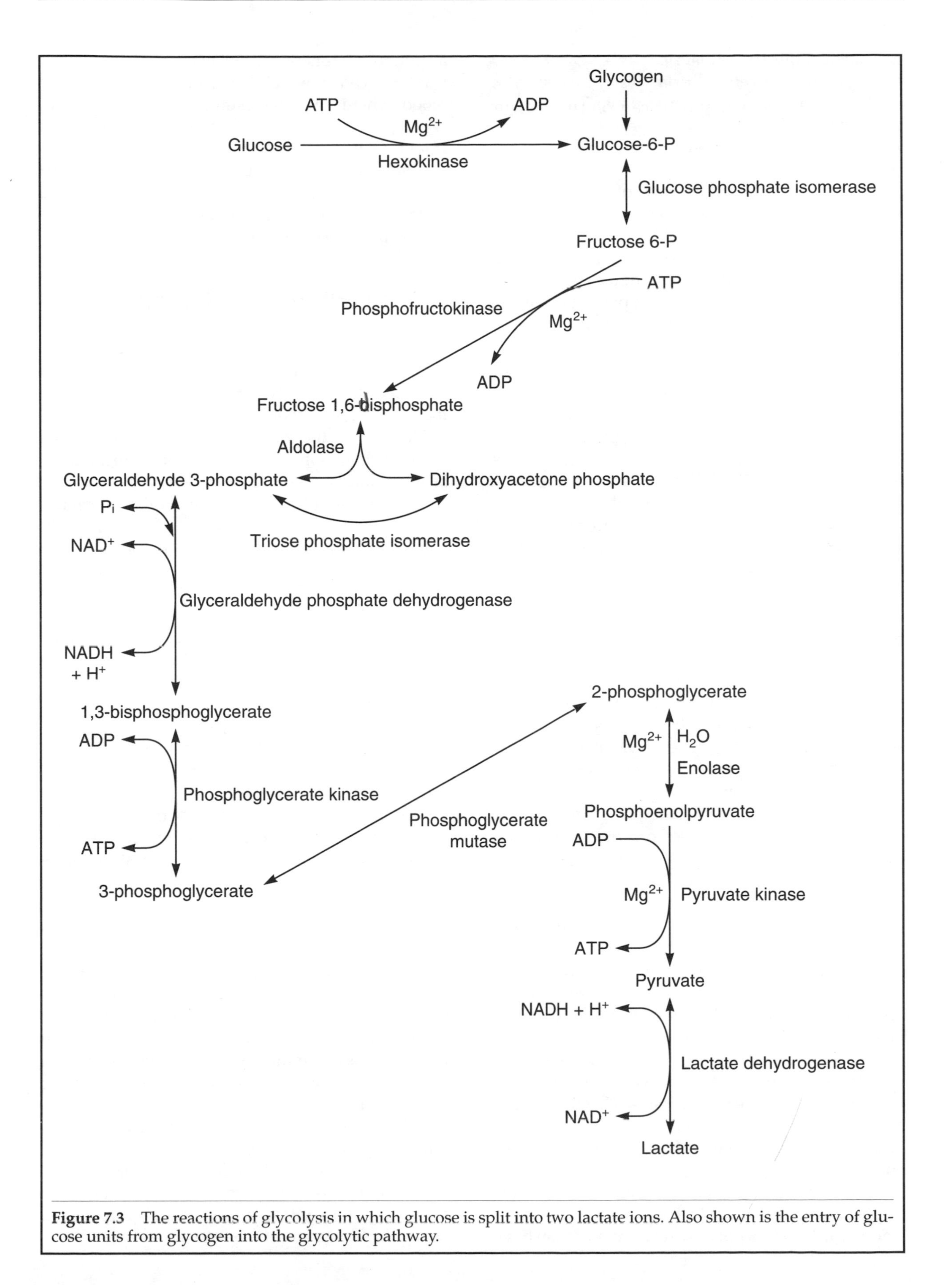

Figure 7.3 The reactions of glycolysis in which glucose is split into two lactate ions. Also shown is the entry of glucose units from glycogen into the glycolytic pathway.

the enzyme triose phosphate isomerase catalyzes the reversible interconversion of dihydroxyacetone phosphate into glyceraldehyde 3-phosphate. This enzyme ensures that all of the carbon atoms in fructose 1,6-bisphosphate are funneled through the glycolytic pathway.

At this point in the pathway, all of the original carbon atoms in each glucose molecule are in the form of two molecules of glyceraldehyde 3-phosphate. Glyceraldehyde phosphate dehydrogenase carries out the following: (a) It oxidizes the aldehyde group of glyceraldehyde to an acid group; (b) it reacts the acid group with a phosphate (Pi) to make an anhydride bond that is energy-rich; and (c) it reduces NAD^+ to NADH + H^+ when the aldehyde is oxidized. A high cytosolic NAD^+/NADH ratio helps drive this reaction.

Phosphoglycerate kinase catalyzes a substrate-level phosphorylation reaction in which an ATP is generated from ADP. Unlike oxidative phosphorylation, which generates ATP by phosphorylating ADP in the mitochondrion while electrons are transferred to oxygen, substrate-level phosphorylation generates an ATP by transferring a phosphate group from a substrate (1,3-bisphosphoglycerate in this case) to ADP. After this reaction, the ATP balance scheme starting from glucose is zero (two ATP used and two ATP generated). Remember, two 1,3-bisphosphoglycerates are obtained from each glucose.

Phosphoglycerate mutase catalyzes the movement of a phosphate group from carbon three to carbon two of the glycerate molecule. Mutases catalyze intramolecular phosphate transfer reactions. Enolase catalyzes the dehydration of 2-phosphoglycerate to form the energy-rich molecule phosphoenolpyruvate. Pyruvate kinase carries out the second substrate-level phosphorylation reaction, generating ATP from ADP and leaving the product pyruvate.

As mentioned, pyruvate has two major fates: reduction to lactate or entry into the mitochondrion for complete oxidation. If the former occurs, the NADH + H^+ generated in the glyceraldehyde phosphate dehydrogenase reaction is oxidized to NAD^+ and the pyruvate is reduced to lactate. If the pyruvate enters the mitochondrion, then the NADH + H^+ generated in the glyceraldehyde phosphate dehydrogenase reaction must be converted back to NAD^+ by one of two shuttle mechanisms. This must occur or glycolysis will come to a halt due to a lack of NAD^+. We will discuss this process in a later chapter.

If the final product of glycolysis is lactate (as opposed to pyruvate, which enters the mitochondrion), the lactate is not a waste product. It can have a number of functions. For example, lactate may leave the cell in which it is formed and enter the blood, where it may be taken up by another type of tissue and oxidized to pyruvate and then to CO_2 and H_2O. Lactate may enter liver cells and be a precursor for making glucose (i.e., gluconeogenesis), or it may remain in the cell where it is formed and be used either as a source of energy or be changed into glycogen by reversal of glycolysis.

Regulation of Glycolysis

Glycolysis is a pathway that can generate ATP. It also plays a major role in providing substrate for the citric acid cycle in the form of pyruvate. If pyruvate is the main fate of (aerobic) glycolysis, it produces a pair of electrons in the form of NADH for the electron transfer chain. Glycolysis does not occur haphazardly; its rate is governed by the energy needs of the cell. In part, regulation depends on the concentration of substrates for the various reactions in the pathway: glucose, glycogen, glucose 6-P, and ADP. The need for glycolysis depends on the rate at which ATP is hydrolyzed to drive endergonic reactions. With ATP hydrolysis, ADP increases in the cytosol and is available for the two substrate-level phosphorylation reactions, the phosphoglycerate kinase and pyruvate kinase reactions. However, the major form of control is in the rate of key reactions.

Normally, regulated enzymes are those that catalyze key irreversible reactions near the beginning of a pathway. There are three irreversible reactions in glycolysis: the hexokinase, the phosphofructokinase, and pyruvate kinase reactions. Of these, only the hexokinase and phosphofructokinase reactions are major points of control, at least for skeletal muscle. In liver, the pyruvate kinase reaction is subject to control, as will be discussed later.

Hexokinase can be regulated by the product of its reaction, glucose 6-P. If the concentration of glucose 6-P increases too much, either because the pathway is being slowed down at another step or because the breakdown of glycogen is producing it at a sufficiently rapid rate, then glucose 6-P binds to a site on the hexokinase enzyme and slows down the reaction rate. We call this process feedback inhibition or product inhibition. As mentioned previously, inhibition of hexokinase results in a sharp decrease in glucose uptake into a cell.

Phosphofructokinase (PFK) is the major regulator of the flux from glucose 6-P to pyruvate and is under complex control. In muscle, PFK activity can be regulated in a variety of ways:

1. PFK can be phosphorylated by several protein kinases, resulting in increased activity. Protein kinases are enzymes that transfer a phosphate group from ATP to hydroxyl groups of serine, threonine, or tyrosine in proteins. This covalent phosphorylation changes the performance of the proteins. The activity of protein kinases is regulated in a number of ways, as will be investigated later.
2. PFK may bind to certain proteins in the cell, which can alter its sensitivity to inhibitors.
3. PFK exists in dimeric or tetrameric forms (i.e., two or four bound subunits); the tetramer is more active than the dimer. In liver, L_4 is the principal active tetramer, while M_4 is the principal active tetramer in muscle.
4. PFK is subject to complex modulation by a variety of positive and negative effectors. In this regard, it is an allosteric enzyme, that is, it has one or more allosteric sites where effector molecules (also known as allosteric effectors) may bind and, in so doing, alter the activity of the active site of the enzyme. Binding of a positive effector will increase activity, whereas binding of a negative allosteric effector will decrease activity. An enzyme can have multiple positive and negative effectors. Also, a substrate may bind to an allosteric site as well as the active site.

ATP, citrate, and H^+ ions are negative effectors. ATP, a substrate which binds to the active site, can also bind to a negative allosteric site and inhibit the activity of the enzyme. Citrate, from the mitochondrion, changes the more active tetrameric form of PFK to the less active dimeric form. When the concentration of citrate increases in the mitochondrion, some of it passes into the cytosol, indicating that mitochondrial oxidative phosphorylation is working sufficiently fast to supply the energy needs of the cell. In this case, it slows down glycolysis by inhibiting PFK. For example, in a rested muscle, fat oxidation can easily provide the low level of ATP regeneration needed. Citrate will slow down glycolysis because this pathway is not needed; glycolysis uses valuable carbohydrate. H^+ ions (i.e., a decrease in pH) act to promote ATP binding to its negative allosteric site to decrease PFK activity.

There are a number of positive effectors of PFK. Fructose 1,6-bisphosphate (abbreviated F $1{,}6P_2$), a product of the PFK reaction, can bind to a positive allosteric site to increase the activity of the enzyme. Fructose 2,6-bisphosphate (F $2{,}6P_2$) is formed from fructose 6-P by an enzyme known as PFK-2. Epinephrine increases the activity of PFK-2, which increases the F $2{,}6P_2$ concentration. Like F $1{,}6P_2$, F $2{,}6P_2$ also binds to an allosteric site to increase the activity of the PFK. F $2{,}6P_2$ is not a significant activator of PFK in muscle but is in liver. ADP, and to a greater extent AMP, can bind to the same allosteric site to increase the activity of PFK. The concentration of these (particularly AMP) increase in hard-working muscle, thus stimulating glycolysis to provide ATP. Pi and NH_4^+ can bind to a positive allosteric site to increase PFK activity. The concentration of Pi increases in muscle in proportion to the decline in CP concentration. NH_4^+ concentration will increase in hard-working muscle due to the action of the enzyme adenylate deaminase, as discussed previously.

The ATP concentration in muscle does not change much unless the muscle is working at a very high rate. Buffering of ATP by creatine phosphate is one of the main reasons for this. However, even a small decrease in ATP results in a larger increase in ADP, and especially AMP, because ATP concentration is so much greater due to the high activity of the enzyme adenylate kinase (also known as AMP kinase or myokinase).

Fates of Pyruvate in Muscle

In addition to being oxidized in the mitochondrion or reduced to lactate in the cytosol, pyruvate can accept an amino group from the amino acid glutamic acid in a transamination reaction. The pyruvate is then changed to the amino acid alanine. Exercising muscle releases alanine to the blood, a topic we will consider in another chapter.

If we focus on the two major fates of pyruvate in muscle, reduction to lactate in the cytosol or pyruvate oxidation in the mitochondrion, we can assess the factors that determine which process predominates. The deciding factors are these:

1. The rate of pyruvate formation or the glycolytic flux rate.
2. The cytosolic redox state, which is the concentration ratio of [NADH]/[NAD^+]. NADH and NAD^+ participate in oxidation reduction reactions as substrates; thus major relative changes in their concentration influence the direction of reversible redox reactions.
3. The number and size of mitochondria, which reflect the potential for oxidative phosphorylation.
4. The availability of oxygen, which is the final acceptor of electrons in oxidative phosphorylation.

5. The total activity of LDH and the isozyme type of LDH, which is important because the greater the activity of LDH and the more the isozyme type M_4 or M_3H, the greater the chance that pyruvate will be reduced to lactate.

If we compare slow twitch and fast twitch skeletal muscle fibers with heart muscle fibers, we learn some important metabolic lessons. First, their typical activity differs greatly. Heart muscle is continuously active because the heart is always working. Slow twitch (ST) muscle fibers are involved in low intensity activity and are also active during strong contractions. These fibers generate the same tension as fast twitch muscle fibers, but they cannot shorten as rapidly. Fast twitch (FT) fibers are usually active during intense activity but are not normally active during low intensity activity. They can shorten faster than slow twitch muscle fibers primarily because they have different myosin isozymes.

Table 7.1 summarizes important differences between the heart muscle and ST and FT skeletal muscle fibers. FT fibers can hydrolyze ATP at a maximum rate that exceeds ST fibers and heart muscle. Thus, the regeneration of ATP must also be faster in FT fibers. However, ATP regeneration in FT fibers will involve more glycolysis and less fuel oxidation compared to the ST and heart muscle fibers because the FT fibers have a greater capacity for glycolysis. In addition, FT fibers have a lower capacity for fuel oxidation due to a lower blood supply and thus less oxygen. They also have fewer mitochondria, where fuel oxidation takes place. In summary, an FT fiber would produce more lactate and consume less oxygen when regenerating ATP compared to an ST fiber or heart muscle.

If a person begins an endurance training program, such as cycling, running, or swimming, significant changes take place in the muscles that participate in the training. The capacity for glycolysis generally does not increase and may even decrease. The number and size of the mitochondria increases. Also, the ability to deliver oxygen to the muscle fibers increases. We should therefore expect that, for the same rate of ATP hydrolysis after training compared to before, less lactate would be produced and more fuel oxidized to generate ATP. Also, the trained muscle is better able to use fat as a fuel, thus using less carbohydrate—a useful consequence of endurance training because the body has far greater fat stores than carbohydrate stores. Furthermore, carbohydrate is saved (less blood glucose used) for other obligatory glucose users, such as the brain.

The formation of lactate by a muscle means that pyruvate is being produced in the cytosol by the glycolytic pathway. It may or may not mean a lack of oxygen because pyruvate formed in glycolysis can either be reduced to lactate or enter a nearby mitochondrion for further oxidation. The path pyruvate takes involves a chance encounter: It meets a mitochondrial membrane or a lactate dehydrogenase enzyme. Since most muscle has a higher capacity for pyruvate reduction, some lactate should be formed whenever glycolysis is even slightly active.

Table 7.1 Relative Comparisons for Heart Muscle (HM) and Slow Twitch (ST) and Fast Twitch (FT) Skeletal Muscle Fibers for a Variety of Metabolic Factors

Factor	Fiber type comparison
Maximum rate of ATP hydrolysis	FT > ST > HM
Maximum glycolytic flux rate	FT > ST > HM
Blood supply or availability of oxygen	HM > ST > FT
Fiber size	FT ≥ ST > HM
Maximum oxidative capacity	HM > ST > FT
Percentage of LDH as the isozyme M_4	FT > ST ≥ HM

Summary

Glycolysis is a metabolic pathway consisting of a series of enzyme-catalyzed reactions in which carbohydrate, in the form of glucose or glucose derived from stored glycogen, is broken down to lactate in cell cytosol. Much of the carbohydrate for glycolysis comes from the diet. Although most dietary carbohydrate is broken down in the intestinal tract to glucose, fructose, and galactose, glucose is of major importance. Glucose is transported from the blood across cell membranes by a glucose transporter that exists in five forms. The GLUT4 transporter is found in muscle and fat cell membranes and its content in these membranes is greatly increased by insulin. Exercise also increases the uptake of glucose into muscle. In cells, the glucose is ultimately broken down in the glycolytic pathway to pyruvate. Pyruvate can be reduced to lactate by the enzyme lactate dehydrogenase, or it can enter the mitochondrion, where it is a fuel for oxidative phosphorylation.

Glycolysis is a carefully regulated process with most of the control exerted by the enzyme phosphofructokinase (PFK). This enzyme is sensitive to the binding of a number of positive and negative modulators to allosteric sites. Rest or low level muscle activity is associated with a low activity of glycolysis, whereas vigorous exercise greatly increases its rate in order to produce ATP. Whether the final product of glycolysis is pyruvate for the mitochondrion or lactate which can appear in the blood depends on a variety of factors. Fast twitch muscle fibers are particularly well-suited to form lactate since they have a higher content of glycolytic enzymes and lower capacity to oxidize lactate. Although glycolysis is an anaerobic process, because lactate formation requires no oxygen, the presence of lactate does not mean that the cell is in an anaerobic state.

Chapter 8

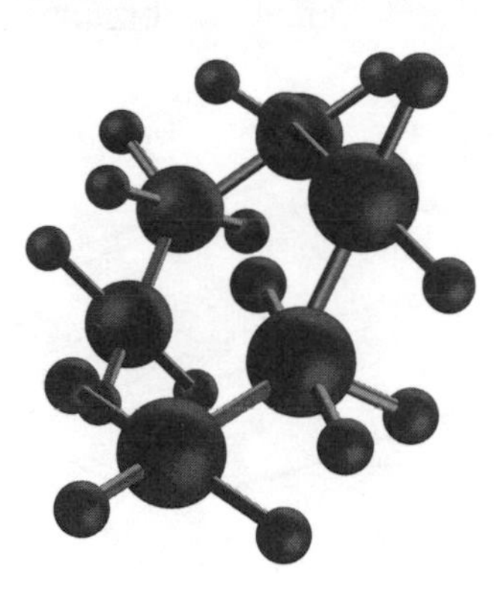

Oxidative Phosphorylation

Oxidative phosphorylation is the formation of ATP from ADP and Pi in association with the transfer of electrons from fuel molecules to coenzymes to oxygen. Although probably the best term to use, oxidative phosphorylation is also known by other names. The biologist may call it cellular respiration, whereas the exercise physiologist may describe it as oxidative metabolism. Whatever the term used, oxidative phosphorylation provides the majority of our ATP—more than 85%, with the remainder arising from substrate-level phosphorylation, such as in glycolysis.

Oxidation is the transfer of electrons from fuels to oxygen. Before reaching oxygen, the electrons are first passed to oxidized coenzymes, reducing these. The electrons are then transferred from the reduced coenzymes to oxygen. Thus the oxygen you breathe in is used to accept electrons that originated in fuel molecules. When reduced, oxygen becomes water. During the electron transfer from reduced coenzymes to oxygen, enough free energy is released to phosphorylate ADP with Pi to make ATP. This process is the phosphorylation or ATP-regeneration part of oxidative phosphorylation. The fuel molecules are mainly carbohydrates, in the form of glucose or glycogen, or fat, in the form of fatty acids and ketone bodies. However, amino acids, minus their amino groups, can be oxidized as well.

Mitochondria

Oxidative phosphorylation takes place in cell mitochondria. Oxygen is delivered to mitochondria by transport in the blood, attached to hemoglobin molecules in red blood cells, and then by diffusion from the small capillaries into the cells. Certain cell types, chiefly type I or ST muscle fibers and heart muscle cells, contain a protein known as myoglobin that helps the diffusion of oxygen. These myoglobin-containing cells have a high capacity for oxidative phosphorylation.

Figure 8.1 illustrates the essential features of a mitochondrion, with its two membranes. The outer membrane is permeable to most ions and molecules up to a molecular weight of 10,000 Da (Daltons). The outer membrane is thus leaky because it contains channels or pores composed of proteins known as porins, through which solutes can pass. The enzyme hexokinase is typically found attached to the cytosolic side of the outer membrane. The inner membrane is impermeable to most ions and polar molecules unless they have specific transporters or carriers. Between the two membranes is the intermembranous space. The inner membrane is formed into bulges called cristae, which greatly increase its surface area. The density of cristae in mitochondria is generally much higher in tissues

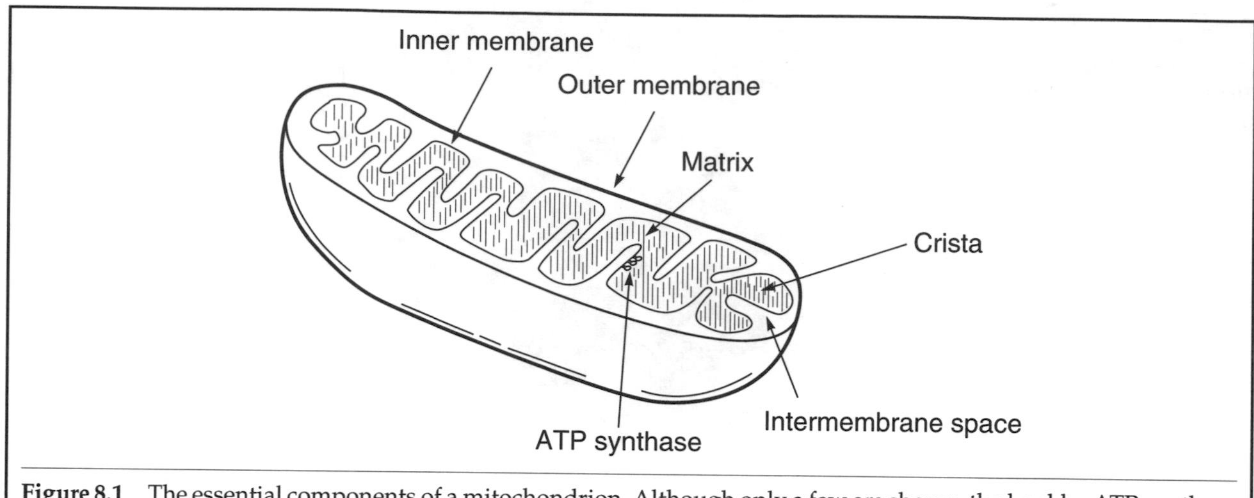

Figure 8.1 The essential components of a mitochondrion. Although only a few are shown, the knobby ATP synthase is found throughout the inner membrane.

where the rate of oxidative phosphorylation is high, such as the heart.

The inner membrane is loaded with enzymes and proteins for transferring electrons to oxygen and the enzyme ATP synthase that converts ADP and Pi to ATP. In Figure 8.1, ATP synthase is shown as small knobs in the mitochondrial inner membrane. In the center of a mitochondrion is the matrix, a viscous medium containing all the enzymes of the citric acid cycle (except succinate dehydrogenase), the enzymes of beta oxidation of fatty acids, other enzymes, and mitochondrial DNA. Mitochondrial DNA codes for 13 of the 700 or so mitochondrial proteins (found in the outer and inner membranes, the intermembrane space, and the matrix)—the remainder are coded by nuclear genes. Following transcription and translation, proteins are imported into the mitochondrion. Mitochondrial DNA, which in humans has 16,659 bp, also codes for two rRNA molecules and 22 tRNA molecules—in sum, a total of 37 genes. Virtually all mitochondrial DNA is maternally contributed; less than 0.1% arises from sperm.

The chemical equations that follow illustrate the complete oxidation of two types of fuels: glucose, a representative carbohydrate, and palmitic acid, a very common fatty acid.

$$C_6H_{12}O_6 + 6\ O_2 \longrightarrow 6\ CO_2 + 6\ H_2O$$

$$C_{16}H_{32}O_2 + 23\ O_2 \longrightarrow 16\ CO_2 + 16\ H_2O$$

These balanced equations describe the oxidation of these fuels in oxidative phosphorylation. These equations also describe the oxidation of these fuels if they were completely broken down in burning.

The respiratory quotient (RQ) is the molar ratio of CO_2 produced divided by the O_2 consumed during fuel oxidation. Using the previous equations, RQ is 6/6 = 1.0 for glucose and 16/23 = 0.7 for palmitic acid. When the oxygen consumed and carbohydrate produced are measured at the mouth of an animal or human, we use the term respiratory exchange ratio or RER instead of RQ, although they represent the same thing.

In the cell, oxidation of glucose and fatty acids is tightly coupled to ADP phosphorylation, or phosphorylation of ADP with Pi to make ATP. Let us rewrite the above equations, but now including the phosphorylation part; that is, we will include the number of moles of ATP formed for each mole of fuel oxidized in the cell.

$$C_6H_{12}O_6 + 6\ O_2 + 36(ADP + Pi) \longrightarrow 6\ CO_2 + 36\ ATP + 42\ H_2O$$

$$C_{16}H_{32}O_2 + 23\ O_2 + 129(ADP + Pi) \longrightarrow 16\ CO_2 + 129\ ATP + 145\ H_2O$$

The number of water molecules generated is increased by 36 for glucose and 129 for palmitic acid compared to the equations that only showed oxidation because when ATP is formed from ADP and Pi, a water molecule results. In contrast, with ATP hydrolysis, a molecule of water is needed to hydrolyze the ATP.

The P/O ratio (sometimes called the ATP/O ratio) is the number of ATP formed for each atom of oxygen consumed. For palmitic acid, P/O is

$$129/(23 \times 2) = 2.8.$$

For glucose, P/O is

$$36/(6 \times 2) = 3.0.$$

Comparing these two numbers reveals that, for the same amount of oxygen consumed, you get more ATP from glucose than you do from a fatty acid.

General Mechanism of Oxidative Phosphorylation

The free energy released during electron transfer from reduced coenzymes (NADH and $FADH_2$) to oxygen gets channeled into the phosphorylation of ADP with Pi to make ATP or drive the reaction.

$$ADP + Pi \longrightarrow ATP + H_2O$$

The mechanism behind oxidative phosphorylation parallels the way electricity is generated by falling water. A dam creates potential energy by raising water up to a high level. When the water falls down through special channels, the kinetic energy rotates turbine blades in a magnetic field, producing an electric current. During electron transfer from NADH and $FADH_2$ to oxygen, free energy is released. This energy is employed to pump protons (H^+) from the matrix side of the inner membrane of the mitochondria to the outside or cytosolic side. An electrochemical gradient results in which the cytosolic side of the inner membrane is more positive in charge (the electro part of the gradient) and has a higher concentration of H^+ (the chemical part of the gradient). When protons (H^+) return down the gradient through a special protein complex, the free energy released is used to make ATP from ADP and Pi. In other words, return of protons down their gradient is harnessed into driving ADP phosphorylation much like the energy of falling water is harnessed to make electricity. Figure 8.2 summarizes this process.

In reactions involving energy-rich phosphates, we describe their standard free energy potential by the $\Delta G^{\circ\prime}$. With electron transfer during oxidative phosphorylation from NADH to oxygen, generating water, the driving force is called a redox potential. Under standard conditions this is described by $\Delta E^{\circ\prime}$ (the standard redox potential). Values for $\Delta E^{\circ\prime}$ are measured in volts (V), based on standard electrode potentials that we will not consider here. For the reduction of an oxygen atom (that is, one half an oxygen molecule) by two electrons producing a water molecule, the $\Delta E^{\circ\prime}$ is 0.82 volts, shown with Equation 1.

1. $1/2\ O_2 + 2\ H^+ + 2\ e^- \longrightarrow H_2O \quad \Delta E^{\circ\prime} = 0.82\ V$

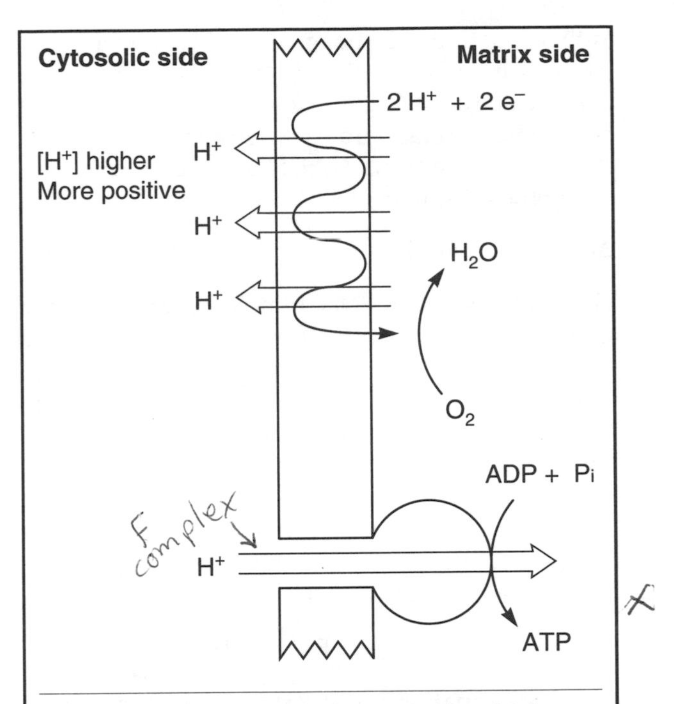

Figure 8.2 A section of the inner mitochondrial membrane showing how electron transfer from substrates through coenzymes to oxygen, forming water, is coupled to ATP formation. During electron transfer, protons (H^+) are pumped across the membrane, creating an electrochemical gradient on the cytosolic side. Their return through ATP synthase drives ATP formation from ADP and inorganic phosphate Pi.

Equation 2 illustrates the reduction of NAD^+ to form NADH + H^+.

2. $NAD + 2\ e^- + H^+ \longrightarrow NADH \quad \Delta E^{\circ\prime} = -0.32\ V$

In oxidative phosphorylation, electrons are transferred from NADH to oxygen, reducing the oxygen to water, and at the same time, NADH is oxidized to NAD^+. We could summarize this process by combining Equations 1 and 2, but we would need to reverse Equation 2 to show it as oxidation, which is what actually happens. The two equations and their algebraic sum (below the line) appear as follows:

$$1/2\ O_2 + 2\ H^+ + 2\ e^- \longrightarrow H_2O \quad \Delta E^{\circ\prime} = 0.82\ V$$

$$NADH \longrightarrow NAD^+ + H^+ + 2\ e^- \quad \Delta E^{\circ\prime} = 0.32\ V$$

$$NADH + 1/2\ O_2 + H^+ \longrightarrow NAD^+ + H_2O$$

$$\Delta E^{\circ\prime} = 1.14\ V$$

We can convert $\Delta E^{\circ\prime}$ values in volts to $\Delta G^{\circ\prime}$ values in kJ/mole by using the following equation:

$$\Delta G^{\circ\prime} = -nF\Delta E^{\circ\prime}$$

In this equation, n is the number of electrons transferred (2 in our example). F is the Faraday constant with a value of 96.5 kJ per mole per volt. We can now determine the standard free energy resulting from electron transfer from NADH to oxygen by substituting these numbers into the equation.

$$\Delta G^{\circ\prime} = -2 \times 96.5 \text{ kJ} \times \text{moles}^{-1} \times \text{volts}^{-1} \times 1.14 \text{ volts}$$

The $\Delta G^{\circ\prime}$ is equal to −220 kJ/mole, a number almost six times greater than the standard free energy for ATP hydrolysis, shown previously. The message from this is that a large amount of free energy is released when a pair of electrons on NADH is transferred to oxygen, reducing the oxygen to water.

When the other reduced coenzyme, $FADH_2$, is oxidized to FAD, and two electrons are transferred to oxygen as shown in the following equation, slightly less free energy is released.

$$FADH_2 + 1/2\, O_2 \longrightarrow FAD + H_2O \quad \Delta E^{\circ\prime} = 1.04 \text{ V}$$

Using the equation to convert to free energy, we get a value for $\Delta G^{\circ\prime}$ of −200 kJ/mole. This value is about 10% less than when NADH is oxidized.

The Electron Transfers

From a functional perspective, the mitochondrial oxidative phosphorylation system consists of five protein-lipid complexes located in the inner membrane of the mitochondrion. Four of the complexes make up the respiratory chain, also known as the electron transfer chain. The fifth complex is the ATP synthase.

In three of the four complexes that make up the electron transfer chain, the free energy released is associated with proton pumping from the matrix to the cytosolic side of the inner membrane, as shown in Figure 8.3. Two electrons are transferred via NAD^+, requiring dehydrogenases from fuel substrates, through a series of electron carriers to oxygen. Three ATP are generated per pair of electrons transferred to oxygen. In addition, two electrons can be transferred from other fuel substrates to oxygen via FAD-containing dehydrogenases. When this occurs, two ATP are generated per pair of electrons transferred.

The sequences of electron flow are nothing more than a series of oxidation-reduction reactions, each of which can be shown as

reduced A + oxidized B ——>
oxidized A + reduced B.

Figure 8.4 illustrates the flow of electrons from two representative substrates through their coenzymes and on to oxygen, using the electron transfer (respiratory) chain. The four complexes in the electron transfer chain are shown with Roman numerals. Let

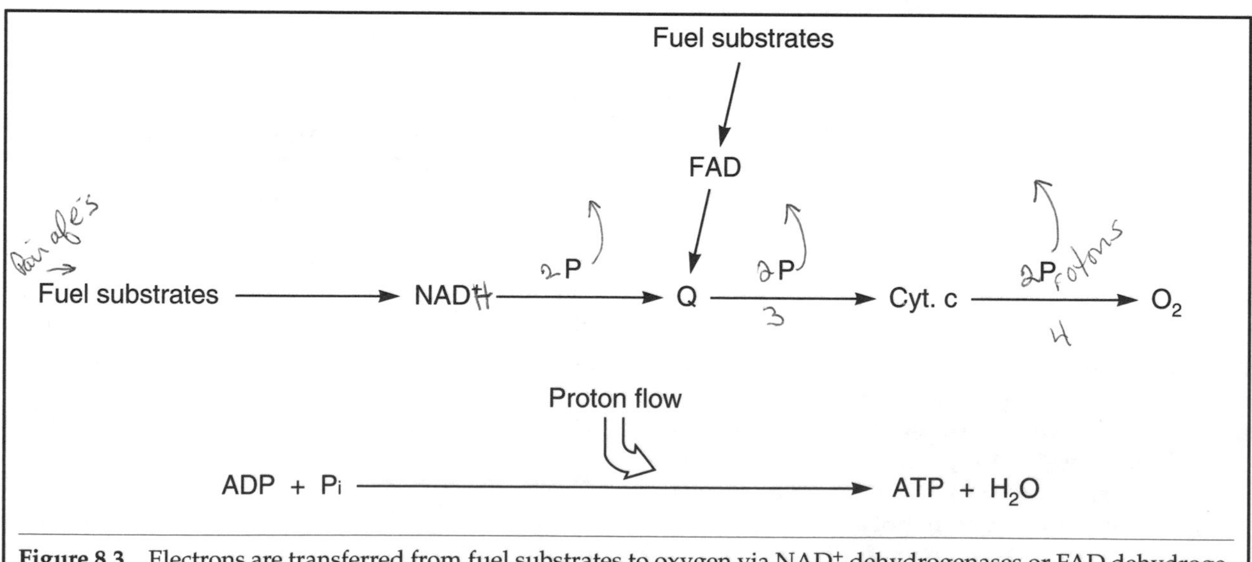

Figure 8.3 Electrons are transferred from fuel substrates to oxygen via NAD^+ dehydrogenases or FAD dehydrogenases. During transfer by NAD^+ dehydrogenases, enough free energy is released to pump protons across the mitochondrial membrane at three sites (P). With FAD dehydrogenases, the lower free energy release pumps protons at only two sites. In the second stage, the proton flow releases free energy, which is harnessed to synthesize ATP from ADP and inorganic phosphate Pi. FAD represents flavin adenine dinucleotide, Q represents coenzyme Q, and cyt. c represents cytochrome c.

us start with the substrate malate. Oxidized by the enzyme malate dehydrogenase (MDH) using NAD^+, it yields oxaloacetate and NADH plus H^+. The two electrons, now on NADH, are passed through the complexes I, III, and IV to oxygen, generating water. Another substrate, succinate, is oxidized by its enzyme succinate dehydrogenase (SDH, a representative of complex II), using FAD and generating $FADH_2$ (the reduced form) and fumarate. The two electrons on $FADH_2$ are then transferred to oxygen, using complexes III and IV. During electron transfers involving complexes I, III, and IV, protons are pumped across the inner mitochondrial membrane, creating the electrochemical gradient. Electron transfer to oxygen from substrates such as succinate involves only two proton pumping complexes (III and IV), whereas electron transfer from substrates to oxygen in which NAD^+ is the coenzyme involve three proton pumping complexes (I, III, and IV); more ATP is thus generated with substrates such as malate.

Complex I: NADH Dehydrogenase

Complex I is a huge complex consisting of at least 30 polypeptides. Its role is to transfer a pair of electrons from NADH in the matrix to coenzyme Q (also known as ubiquinone) in the inner membrane. For this reason, complex I is also known as NADH-Q reductase. Figure 8.5 summarizes the electron transfer reaction of complex I and illustrates the structural changes in coenzyme Q that occur with the acceptance of two electrons and two hydrogen ions. The NADH dehydrogenase complex also contains an FMN-protein (another type of flavoprotein) and an iron-sulphide protein, as intermediates between NADH and Q. During electron transfer from NADH to coenzyme Q, protons are pumped across the inner mitochondrial membrane.

Complex II: Other Flavoprotein Dehydrogenases

Complex II is represented by a number of flavoprotein dehydrogenase enzymes that transfer electrons from substrates to FAD and then to coenzyme Q. An example of an FAD-containing (for flavin adenine dinucleotide) dehydrogenase is succinate dehydrogenase (shown as SDH in Figure 8.4), the only citric acid cycle enzyme located in the inner membrane and not in the matrix of the mitochondrion. We will encounter two more of these FAD-containing complexes, all of which contain a tightly bound coenzyme known as FAD. In the case of succinate, electrons in the form of two hydrogen atoms are transferred from succinate to FAD to make it $FADH_2$. Then the electrons on $FADH_2$ are passed to

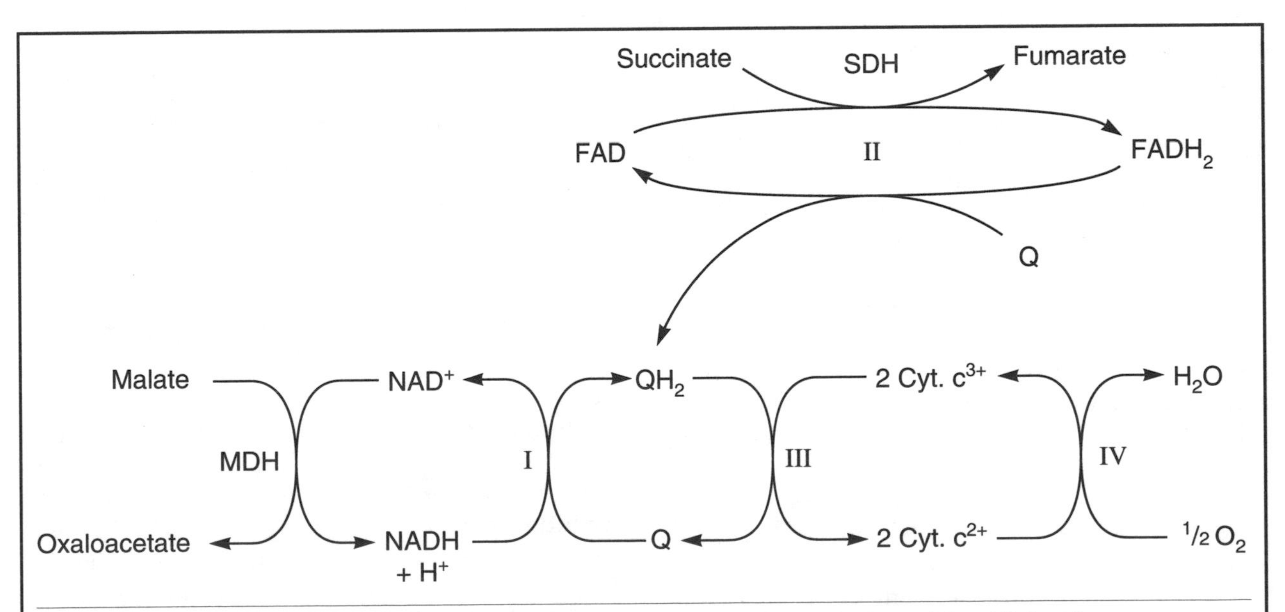

Figure 8.4 Electron transfer from malate and succinate through the four complexes (identified with roman numerals) of the electron transfer chain to oxygen. Proton pumping across the mitochondrial inner membrane involves complexes I, III, and IV. MDH represents malate dehydrogenase, and SDH represents succinate dehydrogenase, a representative member of complex II. FAD and $FADH_2$ represent the oxidized and reduced forms of flavin adenine dinucleotide. Q and QH_2 represent the oxidized and reduced forms of coenzyme Q. Cyt. c represents cytochrome c.

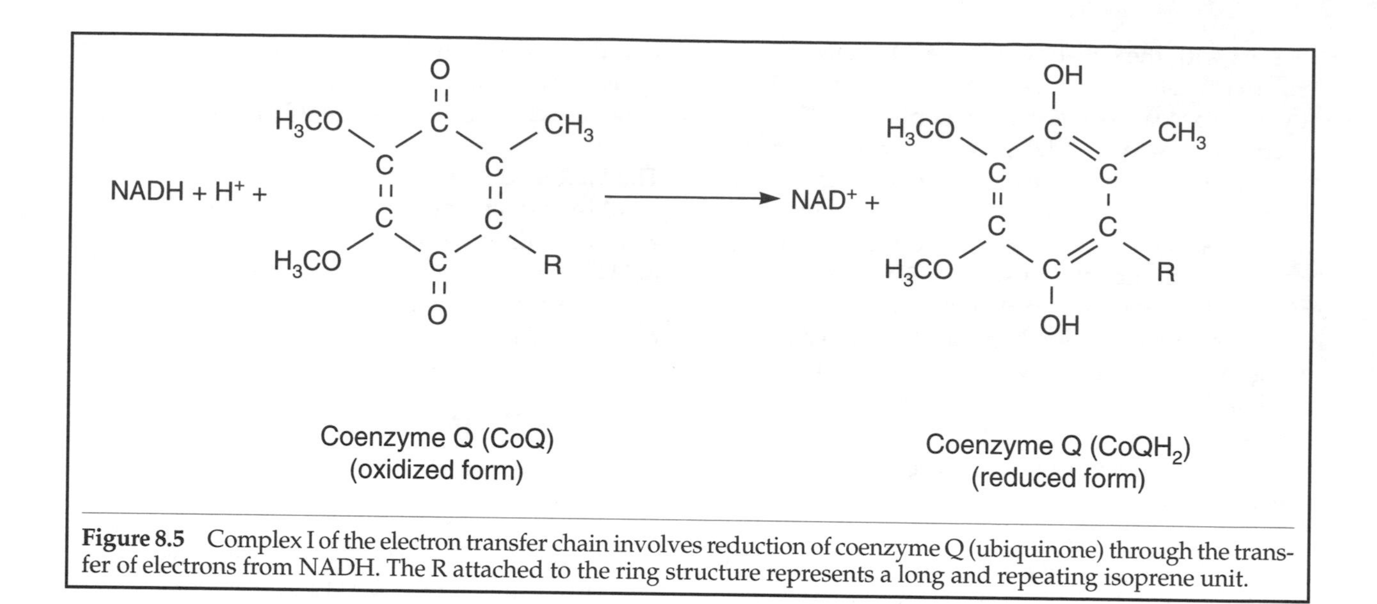

Figure 8.5 Complex I of the electron transfer chain involves reduction of coenzyme Q (ubiquinone) through the transfer of electrons from NADH. The R attached to the ring structure represents a long and repeating isoprene unit.

the oxidized form of coenzyme Q, abbreviated coQ, to make it $coQH_2$. Unlike complexes I, III, and IV, no protons are pumped across the inner membrane during electron transfer during this type of dehydrogenation. Accordingly, when fuel substrates are oxidized by the FAD-containing dehydrogenases, one less ATP molecule is formed (i.e., two instead of three per pair of electrons transferred to oxygen).

Complex III: Cytochrome C Reductase

Complex III transfers electrons from QH_2 to cytochrome c. The cytochromes are a class of heme proteins located in the inner membrane of the mitochondrion. In the center of the heme group is an iron ion that can exist in an oxidized (Fe^{3+}) or reduced state (Fe^{2+}). In complex III, electrons on reduced coenzyme Q (i.e., $coQH_2$) are transferred to cytochrome c, changing the iron from Fe^{3+} to Fe^{2+}. Since reduction of oxidized iron involves accepting only one electron, two cytochrome c molecules must be reduced to accept the electrons from each $coQH_2$. Although complex III involves electron transfer from coenzyme Q to cytochrome c, between coQ and cytochrome c are two other types of cytochromes known as cytochromes b and c_1. During electron transfer from $coQH_2$ to cytochrome c, protons are pumped across the inner mitochondrial membrane. In the following summary equation, note that only the electrons are transferred from $coQH_2$ to cytochrome c. As a result, two protons are left over.

$$coQH_2 + 2 \text{ cytochrome c-}Fe^{3+} \longrightarrow$$
$$coQ + 2H^+ + 2 \text{ cytochrome c-}Fe^{2+}$$

Complex IV: Cytochrome Oxidase

Cytochrome oxidase is a complicated protein that contains two cytochromes, a and a_3, and two copper ions. This complex accepts electrons from reduced cytochrome c and passes them to oxygen, reducing it to water. However, because the oxygen molecule (O_2) contains two atoms of oxygen, we need four electrons to reduce it. We cannot break up the oxygen molecule as suggested earlier by the use of $1/2\,O_2$; it is written this way for convenience. We can summarize the cytochrome oxidase reaction as it really occurs as

$$4 \text{ cytochrome c-}Fe^{2+} + O_2 + 4\,H^+ \longrightarrow$$
$$4 \text{ cytochrome c-}Fe^{3+} + 2\,H_2O.$$

The free energy released during electron transfer in cytochrome oxidase (complex IV) results in proton pumping across the inner membrane.

Coupled Phosphorylation

Before describing how electron transfer is linked to the formation of ATP, we should summarize what has been covered so far. Some fuel substrates are oxidized by transferring two electrons in the form of a hydride ion (H^-) to NAD^+, making it NADH. Then the electrons on NADH are passed through the respiratory chain to oxygen, using complexes I, III, and IV. During electron transfer, proton pumping occurs. Also, there is a large standard free energy change, shown to be −220 kJ/mole.

Other fuel substrates initially transfer electrons to FAD when they get oxidized. These electrons are passed in the form of two hydrogen atoms, each of which contains one electron. The resulting $FADH_2$ then passes the electrons to Q, and then to oxygen. Three complexes are involved: II, III, and IV. However, the standard free energy change is smaller (−200 kJ/mole), and proton pumping occurs only with complexes III and IV.

The proton pumping accompanying electron transfers through the various complexes of the chain creates an electrochemical gradient across the inner mitochondrial membrane. The cytosolic side of the membrane is about 0.5 pH units lower (a higher H^+ concentration) and would have a higher positive electrical charge. This gradient (chemical and electrical, respectively) drives ADP phosphorylation. The concept that electron transport is linked to ATP synthesis by way of proton pumping, conceived by Peter Mitchell in 1961, is known as the chemiosmotic hypothesis, for which Mitchell received a Nobel Prize.

Although the stoichiometry of protons pumped across the membrane is not clearly known, we will assume that four protons (H^+) are pumped at each complex per pair of electrons transferred. The number four is likely for complexes I and III. Therefore, we will assume that 12 H^+ are pumped per pair of electrons transferred from NADH to oxygen and 8 H^+ pumped per pair of electrons transferred from $FADH_2$ to Q to oxygen. This proton pumping generates the potential energy of an electrochemical gradient much like the potential energy of water at the top of the dam creates the kinetic energy of falling water.

ATP synthase or complex V couples proton flow down the gradient into the matrix to phosphorylation of ADP with Pi to make ATP (see Figure 8.6). ATP synthase consists of two parts. The F_0 part, mainly within the inner membrane, acts as a pore to allow protons to pass into the matrix. Tightly associated with F_0 is the F_1 subunit, which bulges into the matrix. It is composed of a number of peptide subunits (not shown in Figure 8.6) and is responsible for combining ADP and Pi into ATP and releasing it into the matrix, although the precise mechanism by which ATP is synthesized is not clear. The substrate for the ATP synthase is $MgADP^-$ and the product is $MgATP^{2-}$, although the role of the magnesium ions is not shown in Figure 8.6.

Electron-transfer-driven proton translocation and ADP phosphorylation are tightly coupled via complex V—the ATP synthase. The flow of only 3 H^+ down the gradient is enough to drive the phosphorylation of ADP with Pi to make one ATP. Protons can also leak across the inner membrane from the cytosolic side to the matrix side without accompanying ATP formation. Such leakage appears to be most prominent in the liver and more important in small mammals. It would allow the proton flow to be dissipated as heat, helping small mammals maintain body temperature.

Mitochondrial Transport of ATP, ADP, and Pi

Most synthesis of ATP occurs in the mitochondria, but most ATP is hydrolyzed in the cytosol. Thus, we

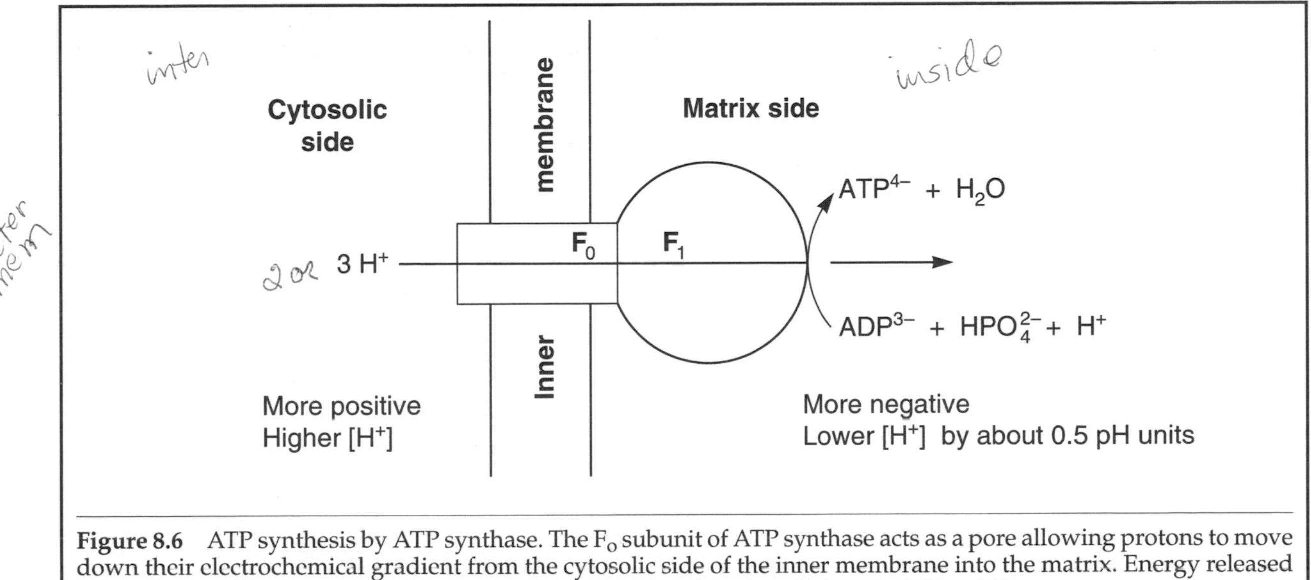

Figure 8.6 ATP synthesis by ATP synthase. The F_0 subunit of ATP synthase acts as a pore allowing protons to move down their electrochemical gradient from the cytosolic side of the inner membrane into the matrix. Energy released during proton flow is used by the F_1 subunit to drive the phosphorylation of ADP to ATP.

must have a way of getting ADP and Pi into the mitochondrion and ATP out of the mitochondrion. ADP and Pi must enter the mitochondrial matrix by crossing the inner membrane, whereas ATP must cross the inner membrane in the opposite direction. Recall that the inner membrane is quite impermeable to most substances, and that polar or charged molecules can only cross if they are transported (translocated) by a specialized carrier protein. Figure 8.7 illustrates how ADP and Pi enter the matrix and how ATP crosses to the cytosolic side. Remember, we ignore the outer mitochondrial membrane because it is so permeable to small molecules.

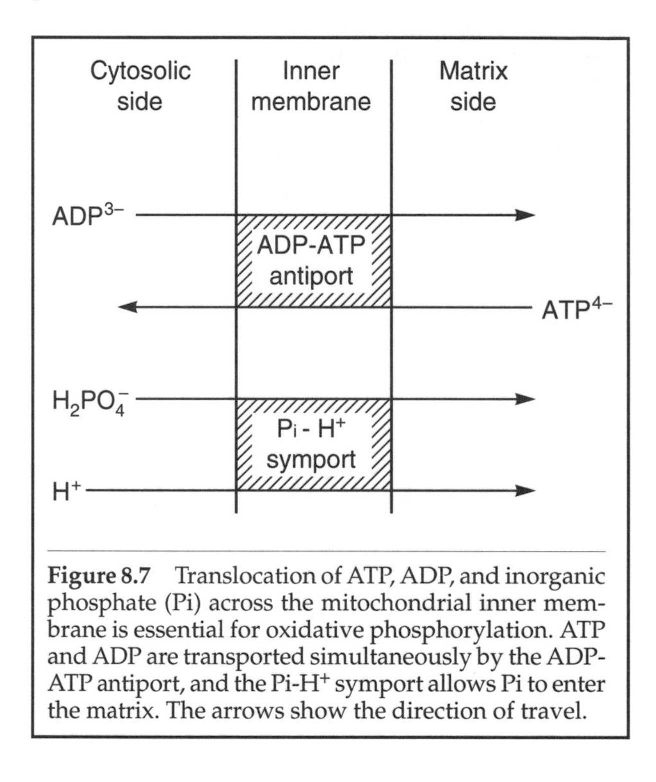

Figure 8.7 Translocation of ATP, ADP, and inorganic phosphate (Pi) across the mitochondrial inner membrane is essential for oxidative phosphorylation. ATP and ADP are transported simultaneously by the ADP-ATP antiport, and the Pi-H^+ symport allows Pi to enter the matrix. The arrows show the direction of travel.

The ADP-ATP antiport transports ADP and ATP. The name antiport means that two substances are crossing a membrane simultaneously in opposite directions, and they are translocated with a carrier protein. By way of contrast, the Pi-H^+ symport transports Pi and H^+ simultaneously across the inner membrane in the same direction. The terms ADP-ATP translocase or Pi-H^+ translocase also describe these carriers. However, the use of antiport or symport is preferable because it gives more information.

The driving force for the ADP-ATP antiport comes from the charge difference between the ADP and ATP. Thus the ATP^{4-} is moving from a more negative region to a more positive region across the inner membrane, or down a gradient, which allows the translocation to proceed with a free energy release. The movement of ATP^{4-} out of the matrix is coupled to moving an ADP^{3-} into it.

The Pi-H^+ symport is driven by the movement of the hydrogen ion (proton) down its concentration gradient, from the cytosolic side of the membrane where its concentration is higher to the matrix where its concentration is lower. This movement is accompanied by a free energy release that drives Pi into the matrix.

Recall that we are assuming four H^+ are pumped for each pair of electrons transported for each of the complexes I, III, and IV. We also assume only three protons are needed to flow down the gradient through the ATP synthase to provide enough energy to phosphorylate one ADP with Pi. The extra proton is lost when each Pi enters the matrix (crosses the inner membrane). Protons, thus, balance out, as the following summary shows.

When a pair of electrons is transferred from NADH to oxygen through the electron transfer chain using complexes I, III, and IV, 12 protons are pumped from the matrix side to the cytosolic side, creating the electrochemical gradient. This results in the formation of three ATP from three ADP and three Pi. Nine protons are needed to flow down the gradient through the ATP synthase to provide the free energy needed to phosphorylate three ADP with three Pi. Three protons are lost when three Pi are transported into the matrix.

Summary

Before leaving this chapter, let us summarize what has been presented.

- Two electrons are always transferred from fuel substrate molecules to coenzymes (NAD^+ or FAD) and from the reduced coenzymes (NADH and $FADH_2$) through the electron transport chain (also known as the respiratory chain) to oxygen.
- When oxygen accepts electrons, it is reduced to water. However, it takes four electrons to reduce one oxygen molecule (containing two atoms of oxygen) to two molecules of water. The reason we breathe oxygen is to use it as an acceptor of electrons in the respiratory (electron transfer) chain.
- Substrates such as malate transfer electrons to NAD^+ to make NADH + H^+ and oxaloacetate (we will see this reaction in the next chapter). Then the electrons on NADH are transferred to oxygen in the electron transport chain. Substrates such as succinate transfer electrons to FAD to make $FADH_2$ and fumarate (we will also see this reaction in the next chapter). This is part of complex II. The electrons on $FADH_2$ are then transferred to Q and then to oxygen in the electron transport chain.
- During the process of electron transfer from NADH to oxygen, free energy is released and then captured by pumping protons across the inner membrane from the matrix to the cytosolic side, creating an electrochemical gradient. During the process of electron transfer from substrates such as succinate to FAD to Q to oxygen, free energy is also released that is captured by pumping protons across the inner membrane.
- The free energy released during the return of protons down their gradient through complex V, the ATP synthase, results in the phosphorylation of ADP with Pi to make ATP.
- The transfer of electrons from substrates (e.g., malate and succinate) to coenzymes (NAD^+ and FAD) and then to oxygen and the phosphorylation of ADP to make ATP are tightly coupled.
- The stoichiometry for electron transfer from NADH to oxygen is as follows: For each two electrons transferred to oxygen, 12 protons (H^+) are pumped across the inner membrane. The return of three protons probably provides enough free energy to produce one ATP from ADP and Pi. For each Pi transferred into the matrix, across the inner membrane, one proton is lost from the gradient. Therefore, we get three ATP formed per pair of electrons transferred to oxygen.
- For electron transfer from substrates such as succinate to FAD and then to oxygen, the stoichiometry is as follows: For each pair of electrons transferred, eight protons are pumped across the inner membrane. Two protons are lost in transporting two Pi into the matrix. Therefore, we get two ATP formed per pair of electrons transferred to oxygen from succinate.

Chapter 9

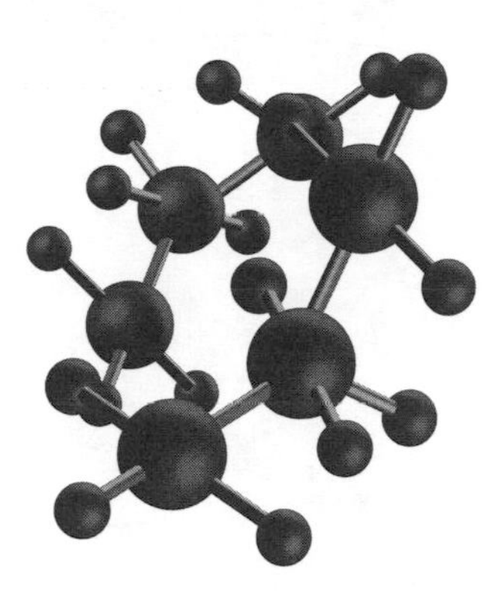

The Citric Acid Cycle

The citric acid cycle, abbreviated CAC, is also known as the tricarboxylic acid (TCA) or Krebs Cycle. The prime function of the CAC is to completely oxidize acetyl groups in a way that will result in ATP formation. The CAC removes electrons from acetyl groups and attaches them to NAD^+ and FAD where they will feed into the electron transport chain. The carbon atoms in the acetyl group are released as carbon dioxide. Each kind of fuel is converted to acetyl groups, attached to coenzyme A (CoA) (see Figure 9.1).

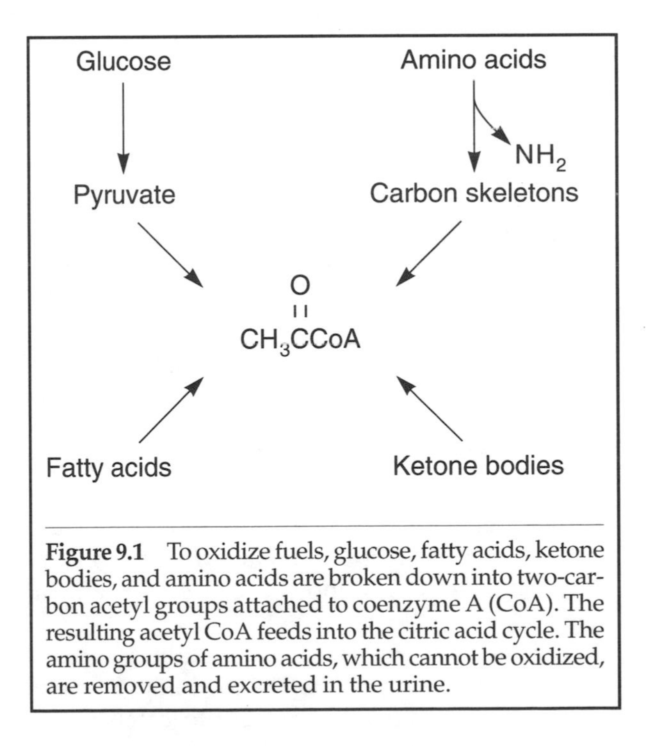

Figure 9.1 To oxidize fuels, glucose, fatty acids, ketone bodies, and amino acids are broken down into two-carbon acetyl groups attached to coenzyme A (CoA). The resulting acetyl CoA feeds into the citric acid cycle. The amino groups of amino acids, which cannot be oxidized, are removed and excreted in the urine.

CoA is derived from pantothenic acid, a B vitamin. It acts as a handle to attach to a number of acyl groups, some of which we will see later. CoA has a terminal SH (sulfhydryl) group to which the acetyl group is attached, forming an energy-rich thioester bond. CoA and acetyl CoA are often written as CoASH and $CH_3COSCoA$, respectively. However, for simplicity, we will use CoA and acetyl CoA.

Oxidation of acetyl CoA accounts for about two thirds of the ATP formation and oxygen consumption in mammals. Acetyl groups enter the CAC where their two carbon atoms appear as CO_2, while the hydrogens and their associated electrons are removed. Figure 9.2 summarizes and simplifies this process. The $:H^- + H^+$ represents the hydride ion and proton that are removed from many fuel substrates. The hydride with its two electrons, shown as dots, gets attached to NAD^+, forming NADH. The two H· represent two hydrogen atoms removed from succinate, although succinate is not identified in the figure. The two hydrogen atoms become attached to FAD forming $FADH_2$.

Figure 9.2 shows that one turn of the CAC consumes one acetyl group and produces four pairs of

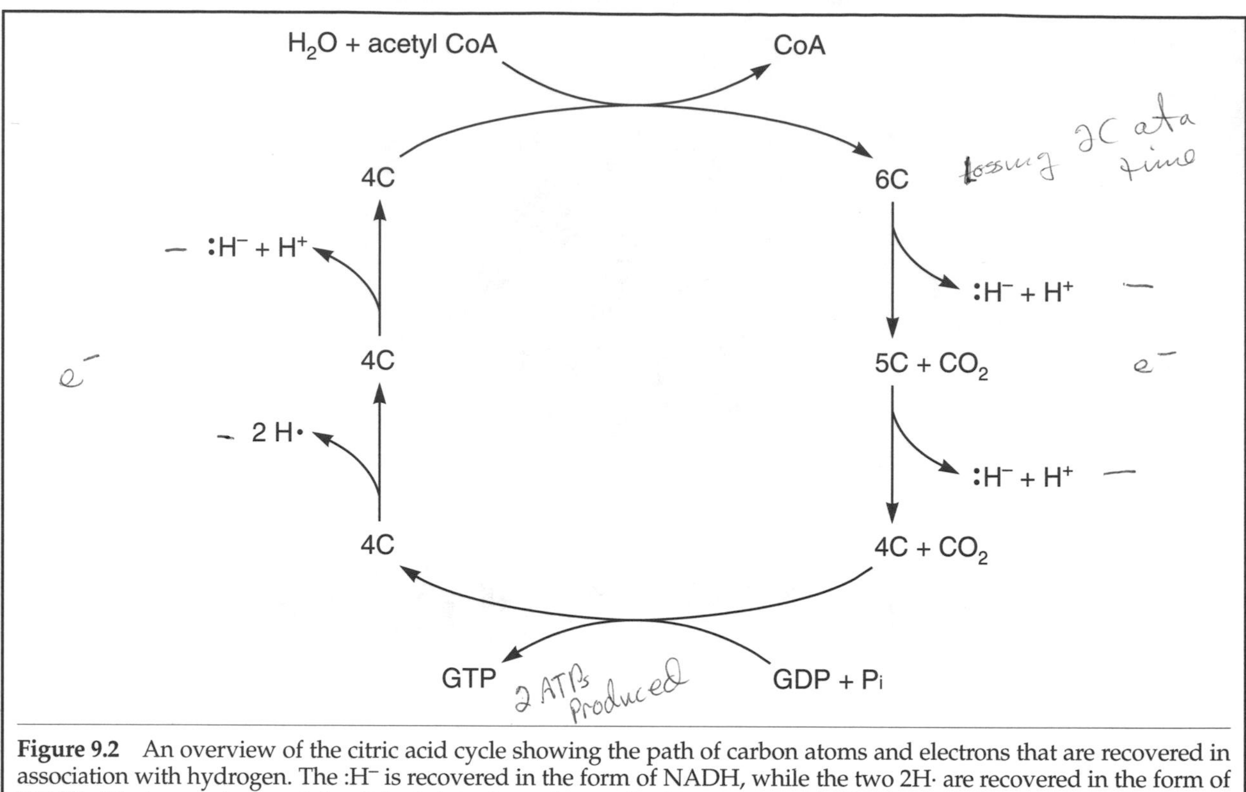

Figure 9.2 An overview of the citric acid cycle showing the path of carbon atoms and electrons that are recovered in association with hydrogen. The $:H^-$ is recovered in the form of NADH, while the two 2H· are recovered in the form of $FADH_2$. The two-carbon acetyl unit feeds into the cycle, attached to coenzyme A (CoA). The carbon atoms are recovered as carbon dioxide (CO_2). During one step in the cycle, enough free energy is released to phosphorylate GDP to make GTP.

electrons, one GTP, and two CO_2. Two water molecules are also consumed, although in the simplified diagram in Figure 9.2 only one H_2O is shown. The GTP produced is an example of substrate-level phosphorylation; that is, formation of an energy-rich phosphate without using oxidative phosphorylation. Remember, ATP produced in glycolysis also occurs via substrate-level phosphorylation.

Reactions of the Citric Acid Cycle

Figure 9.3 shows the complete citric acid cycle including the structures of intermediates. Figure 9.4 summarizes the CAC, without the use of chemical structures.

Hydrolysis of acetyl CoA to acetate and CoA has a $\Delta G^{\circ\prime}$ of -36 kJ/mole. Thus the combining of acetyl CoA and oxaloacetate to form citrate, catalyzed by citrate synthase is virtually irreversible. In the next step, the tertiary alcohol group on citrate is converted to a secondary alcohol group by two steps in the reaction catalyzed by the enzyme aconitase. The first step is a dehydration reaction, in which the tertiary OH group is removed, producing cis-aconitate, which remains attached to the aconitase enzyme. A hydration reaction follows that results in the formation of isocitrate. Notice that the only difference between citrate and isocitrate is the position and type of OH group—tertiary alcohol on citrate and secondary alcohol on isocitrate.

Isocitrate undergoes oxidative decarboxylation, catalyzed by isocitrate dehydrogenase to form a α-ketoglutarate. First oxidation generates NADH and H^+ and then decarboxylation forms CO_2. The decarboxylation spontaneously follows the oxidation but needs the cofactor Mg^{2+}. At this point in the cycle, one of the two carbon atoms on the acetyl group is removed as carbon dioxide.

In the next step, an α-ketoglutarate (also known as 2-oxoglutarate) undergoes oxidative decarboxylation to succinyl CoA. The first step is decarboxylation, followed by oxidation, generating NADH and H^+. The enzyme α-ketoglutarate dehydrogenase catalyzes the same kind of reaction as pyruvate dehydrogenase, a reaction we will encounter soon, and contains the same kinds of subunits and coenzymes. Both enzymes contain three types of polypeptide subunits and five coenzymes. The coenzymes are NAD^+ and CoA, which are loosely bound, and TPP, lipoic acid, and FAD, which are tightly bound and not seen in the simple way the reaction

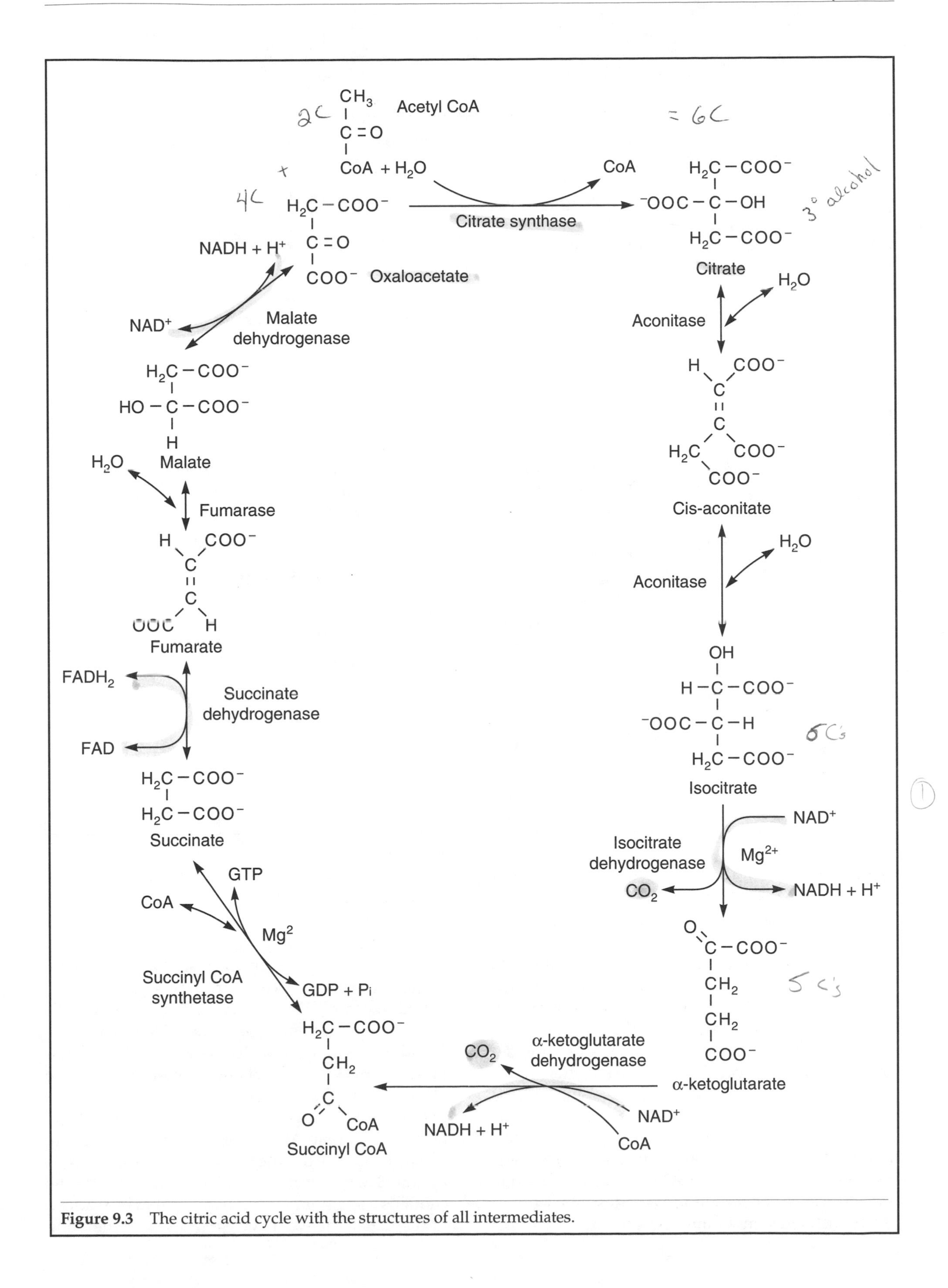

Figure 9.3 The citric acid cycle with the structures of all intermediates.

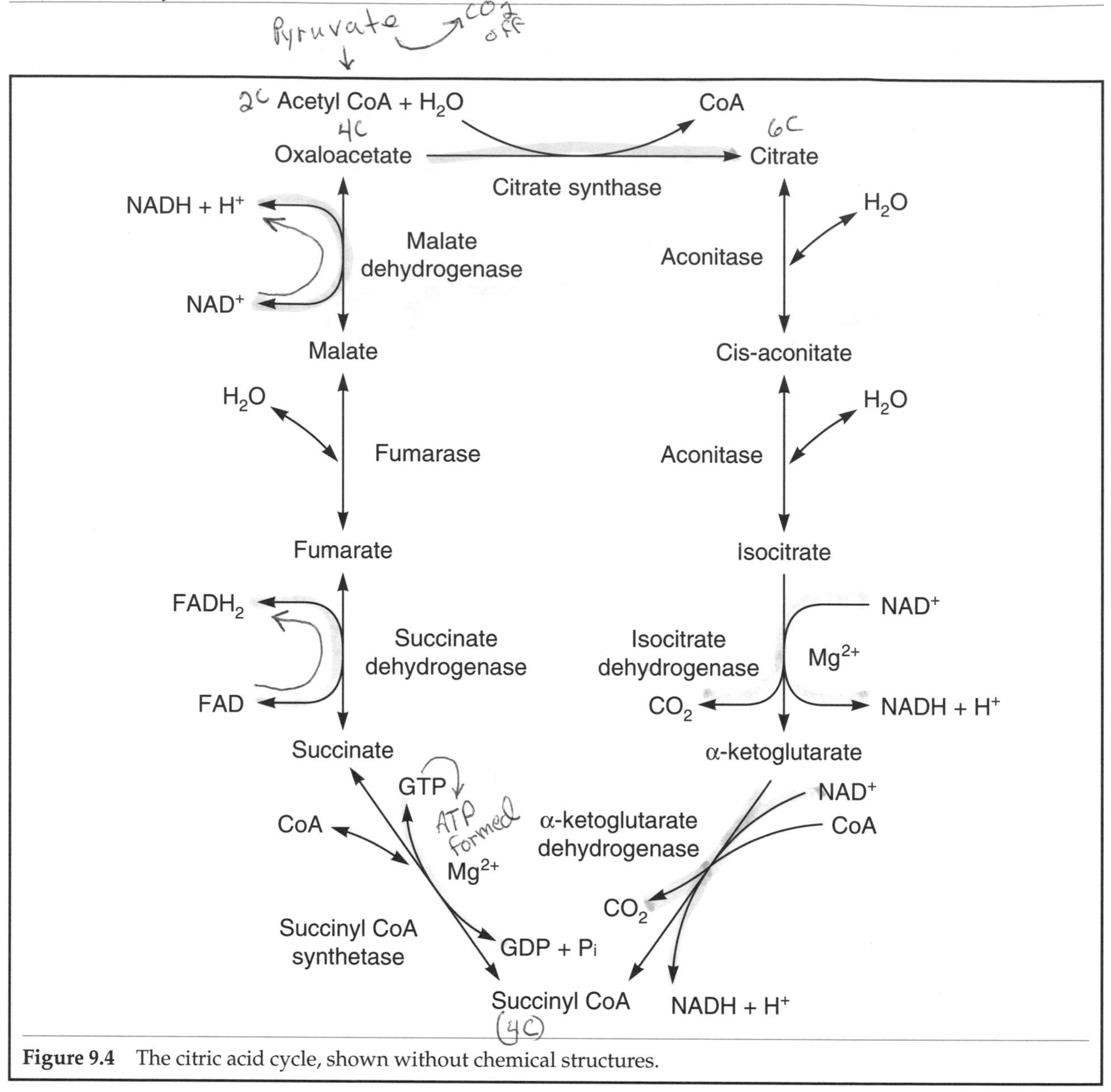

Figure 9.4 The citric acid cycle, shown without chemical structures.

is presented in Figure 9.3 and 9.4. During the α-ketoglutarate dehydrogenase reaction, enough free energy is released to generate the energy-rich succinyl CoA.

In the next step, succinyl CoA is broken down to succinate and CoA in a reaction catalyzed by succinyl CoA synthetase. When this occurs, the free energy released drives the substrate-level phosphorylation of GDP to make GTP. The succinyl CoA synthetase reaction is freely reversible, and the enzyme's name describes the backward reaction, in keeping with the naming of similar reactions in biochemistry. The GTP produced may be used (a) for peptide bond formation during the process of translation, (b) to phosphorylate ADP to make ATP using the enzyme nucleoside diphosphate kinase, described previously, or (c) in signal transduction.

Once succinate is formed, the remaining three reactions of the CAC regenerate oxaloacetate (one of the starting substances of the CAC, along with acetyl CoA) and generate electrons for the electron transport chain (using the enzymes succinate dehydrogenase and malate dehydrogenase). Succinate dehydrogenase (SDH) contains a tightly bound FAD. Unlike the other enzymes of the CAC, which are found in the mitochondrial matrix, SDH is a component of the inner mitochondrial membrane. In the SDH reaction, electrons are transferred from succinate to FAD and then to coenzyme Q. Recall that the SDH reaction is complex II in the electron transport chain.

The product from the SDH reaction, fumarate, is hydrated to malate by the enzyme fumarase. Malate contains a secondary alcohol group that is oxidized in the malate dehydrogenase reaction, generating

NADH + H^+ and oxaloacetate, a starting substrate for a new round of the cycle.

The Complete Cycle

If we add algebraically all the reactions of the CAC we get

$$\text{acetyl CoA} + 3\,NAD^+ + FAD + GDP + Pi + 2\,H_2O \longrightarrow 2\,CO_2 + GTP + 3\,NADH + 3\,H^+ + FADH_2 + CoA.$$

As seen, the CAC does not involve the net production or consumption of oxaloacetate or any other constituent of the cycle and the only thing consumed is an acetyl group and two water molecules.

The reduced coenzymes produced in the CAC (NADH and $FADH_2$) are oxidized in the electron transport chain and their electrons transferred to oxygen. We can show this electron transfer as follows:

$$3\,NADH + 3\,H^+ + FADH_2 + 2\,O_2 \longrightarrow 3\,NAD^+ + FAD + 4\,H_2O$$

Associated with the transfer of electrons from the reduced coenzymes to oxygen, are the tightly coupled ADP phosphorylation reactions, producing ATP. The transfer of electrons from three NADH to oxygen will yield nine ATP. The transfer of electrons from one $FADH_2$ to O_2 will generate two ATP.

In counting the energy-rich phosphates, one must include the GTP formed during the succinyl CoA synthetase reaction. In summary, the complete oxidation of one acetyl group is associated with the formation of 12 ATP. Figure 9.5 illustrates the close coupling of the CAC, the electron transfer chain, and ADP phosphorylation.

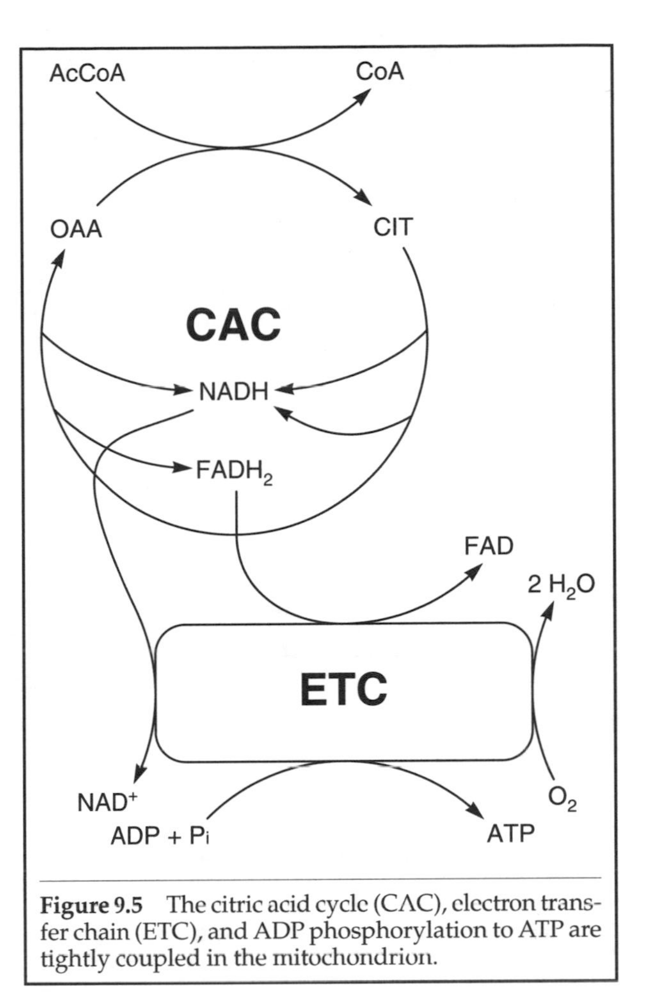

Figure 9.5 The citric acid cycle (CAC), electron transfer chain (ETC), and ADP phosphorylation to ATP are tightly coupled in the mitochondrion.

Regulation of Oxidative Phosphorylation

Because oxidative phosphorylation generates most of the energy for the cell in the form of ATP, its rate should be precisely connected to the rate of ATP hydrolysis. The citric acid cycle is one component of oxidative phosphorylation. As shown in Figure 9.5, the CAC and electron transport chain are tightly linked together because the CAC is the major producer of reduced coenzymes needed to funnel electrons into the respiratory chain.

If we neglect transfer of electrons from $FADH_2$ to oxygen, a single equation can represent oxidative phosphorylation, in which electrons are transferred to a complete oxygen molecule and ADP is phosphorylated with Pi to make ATP:

$$6\,ADP + 6\,Pi + 2\,NADH + 2\,H^+ + O_2 \longrightarrow 2\,NAD^+ + 6\,ATP + 8\,H_2O$$

Two of the eight molecules of water are generated by reduction of a molecule of oxygen and six come about when six ATP are formed.

To understand the regulation of oxidative phosphorylation, we must ask which of the substrates on the left side of the equation (i.e., ADP, Pi, NADH, and O_2) actually limit the overall process. Our discussion will focus on muscle because it has an enormous range of metabolic rate, from complete rest to the vigorous contractions of sprinting. Figure 9.6 aids this discussion. First we must ask, where do the four potential limiting substrates for oxidative phosphorylation come from?

ADP and Pi are formed mainly in the cytosol when ATP is hydrolyzed to drive endergonic reactions. Pi can also increase in the cytosol whenever the concentration of creatine phosphate decreases; that is, CP decreases and free creatine (C) and Pi

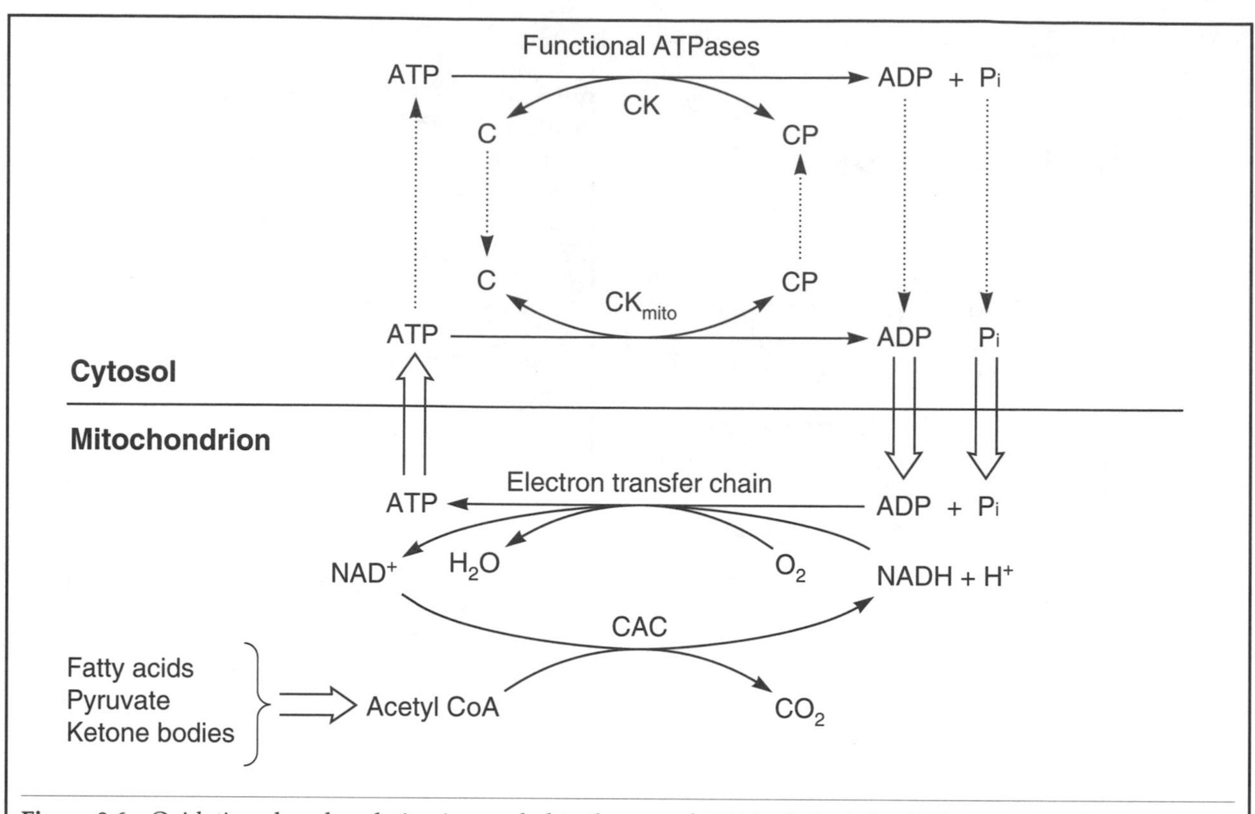

Figure 9.6 Oxidative phosphorylation is coupled to the rate of ATP hydrolysis by ATPases in muscle cytosol. The products of ATP hydrolysis, ADP and inorganic phosphate, Pi, are substrates for oxidative phosphorylation in the mitochondrion, along with electrons from acetyl CoA, and the final electron acceptor, oxygen. ADP may diffuse to the mitochondrion, and ATP diffuse back to the site of ATP hydrolysis, as indicated by the dotted lines. Also, creatine (C) can stimulate ADP formation at the inner membrane by the mitochondrial creatine kinase (CK_{mito}) reaction. Creatine phosphate (CP) is a source of ATP at the site of the ATPases using creatine kinase located there. The movement of C and CP in the energy cycle is called the creatine phosphate shuttle. Double arrows refer to membrane transport. No distinction is shown between the matrix and inner membrane of the mitochondrion.

increase. NADH comes from the CAC, β oxidation of fatty acids in the matrix of the mitochondrion, and the cytosol if pyruvate is not reduced to lactate during glycolysis. NADH is a potential limiting factor because adequate NADH for oxidative phosphorylation requires available substrates for the dehydrogenase reactions that generate it as well as sufficient activity of the dehydrogenase enzymes to drive the NADH-forming reactions.

Oxygen is taken into the lungs when you breathe, diffuses to hemoglobin molecules in red blood cells (i.e., erythrocytes) in the capillaries in the lung, is pumped throughout the body from the heart, and is unloaded from hemoglobin molecules in the capillaries that reach all parts of the body. Oxygen is delivered by way of diffusion from the small capillaries, across the cell membrane, and then to the mitochondrion.

Figure 9.6 shows that ATP is hydrolyzed to ADP and Pi by the functional ATPases, whereas ATP is primarily regenerated via oxidative phosphorylation in the mitochondrion. For tightly coupled ATP hydrolysis and ATP regeneration to happen, ADP and Pi must cross from the cytosol into the matrix and ATP must cross from the matrix to the cytosol to be used again. Transport of ADP, Pi, and ATP across the inner mitochondrial membrane is aided by specific carriers, described in the previous chapter. It is widely believed that, in skeletal and cardiac muscle, much of the cytoplasmic transport of ADP and ATP occurs via C and CP, respectively.

ATP is hydrolyzed in the cytosol of muscle at three major sites: (a) where myosin interacts with actin (the actin-activated ATPase discussed previously), (b) when calcium ions are pumped back into the sarcoplasmic reticulum (the sarcoplasmic reticulum ATPase) and (c) when sodium ions are pumped out of the cell and potassium ions are pumped back in (the sodium-potassium ATPase). At the site of these ATPases is the creatine kinase enzyme (CK) and CP. CK catalyzes the phosphorylation of ADP to make ATP, utilizing CP and producing C. At this level, the net direction of the CK reaction is toward ATP formation. Next, C can diffuse to the outer side of the inner membrane where it is phosphorylated by an ATP, producing CP and ADP. A mitochondrial creatine kinase (CK_{mito}) catalyzes this reaction. The net direction of the CK_{mito} reaction is toward CP. The resulting CP can then diffuse back to the site of the ATPases to rephosphorylate ADP. This process is called the creatine phosphate shuttle.

Use of this shuttle does not prevent ADP from diffusing to the mitochondrion and ATP from diffusing back (see Figure 9.6); it just means that C and CP can carry out the same process. The ATP used to phosphorylate C at the mitochondrial level, using the mitochondrial form of CK, crosses from the matrix at the same time ADP crosses into the matrix (using the ADP-ATP antiport). The ADP-ATP antiport is the most abundant protein in the inner mitochondrial membrane and is in physical association with mitochondrial creatine kinase, located on the cytosolic face of the inner membrane. ATP is formed via oxidative phosphorylation when electrons are transferred from fuel substrates (pyruvate, fatty acids, ketone bodies, and acetyl CoA) to oxidized coenzymes (such as NAD^+) to make NADH. The electrons on NADH are transferred to oxygen using the electron transfer chain.

Figure 9.6 shows where all the limiting players (NADH, ADP, Pi, and O_2) come from. Now let us look at some conditions in skeletal muscle to see what limits the rate of oxidative phosphorylation.

Rested Muscle

Oxygen is readily available in rested muscle; therefore it cannot be limiting. The matrix concentration of Pi is also high enough to sustain a high rate of oxidative phosphorylation and is therefore not limiting. The mitochondrial redox state (i.e., the concentration ratio of $NADH/NAD^+$) is highly reduced, so plenty of NADH is available. The availability of ADP limits oxidative phosphorylation (and thus also the CAC). This means the rate of entry of ADP into the mitochondria, and thus the matrix concentration of ADP, is the most likely candidate to regulate oxidative phosphorylation at rest. Of course, the availability of ADP depends on the rate of ATP hydrolysis in the cytosol. Many studies support this fact. For example, if mitochondria are isolated and placed in a well-oxygenated medium, the rate of oxygen consumption (used as an index of the rate of oxidative phosphorylation) is low. If either ADP or C is added, the rate of oxygen utilization greatly increases. We have already discussed how addition of ADP stimulates oxidative phosphorylation. However, addition of C is also effective by stimulating the mitochondrial creatine kinase enzyme, generating ADP and CP. The ADP then enters the matrix to stimulate oxidative phosphorylation.

Contracting Muscle

Suppose that our rested muscle starts contracting. Of course this means that the rate of ATP hydrolysis greatly increases, and the rate of formation of ADP increases in proportion. The ADP concentration will not rise very much at the site of ATP hydrolysis because of the creatine kinase reaction. However, C concentration will rise, and C will diffuse to the mitochondrion where it will generate ADP and CP, using the mitochondrial creatine kinase. The ADP thus formed will enter the mitochondrial matrix because the ADP-ATP antiport and the mitochondrial creatine kinase are close together. The ADP in the mitochondrion will stimulate oxidative phosphorylation and thus oxygen utilization. As each ADP enters the matrix, an ATP exits because of the ADP-ATP antiport. Intramitochondrial ADP can be increased but only slightly for light to moderate contractions because ADP limits oxidative phosphorylation under these conditions. In other words, we would expect the rate of oxidative phosphorylation to increase in proportion to the entry of ADP into the matrix for all but the most intense dynamic contractions. As the rate of entry of ADP into the mitochondrial matrix increases, it stimulates its own phosphorylation by Pi. ADP also stimulates the rate of electron transfer from NADH to oxygen, thereby increasing the rate of oxygen utilization and the formation of NAD^+. The increase in NAD^+ stimulates the rate of the citric acid cycle because NAD^+ is a substrate for three CAC reactions.

Normally, in rested muscle or in muscle doing most forms of easy to moderate work, ADP is the limiting factor for oxidative phosphorylation—we could call this acceptor control by ADP. It is possible for other factors to also limit oxidative phosphorylation in muscle. For example, at high altitude, where the oxygen content of the air is low, oxygen is limiting. Oxygen could also be limiting during isometric contractions when intramuscular pressure builds up enough to cut off the blood flow and therefore the supply of oxygen to mitochondria. Cross your right foot over your left foot and attempt to extend your left leg while resisting with your right leg. ATP is being hydrolyzed, but because of an increase in intramuscular pressure, blood flow is greatly reduced. In situations of reduced mitochondrial oxygen, glycolysis becomes extremely important as a source of ATP.

NADH can also limit oxidative phosphorylation in muscle. For example, during a long marathon run or bike ride, the availability of carbohydrate to muscle (i.e., muscle glycogen and/or blood glucose) is greatly attenuated. In this case, muscle must rely on oxidation of fatty acids. However, the rate of NADH generation in muscle mitochondria when almost pure fatty acids provide the fuel substrate is limited, making NADH a limiting factor. With endurance training, however, the content of the mitochondrial enzymes increases quite dramatically, and many of these enzymes generate NADH (or $FADH_2$).

This discussion has focused on what regulates oxidative phosphorylation in skeletal muscle. However, the arguments do not necessarily hold for other tissues. For example in the heart, the changes in cytosolic ADP concentration over a fairly wide range of heart rate are very small. This fact could preclude ADP as the primary regulator of oxidative phosphorylation in the heart. In this case, changes in creatine phosphate concentration, matrix NADH concentration, or other factors may combine to regulate the rate of cardiac oxidative phosphorylation to match cellular needs.

Regulation of the Citric Acid Cycle

We can appreciate the control of the citric acid cycle from two perspectives. It can be regulated in a general way because of intimate connections with the overall process of oxidative phosphorylation. In addition, some of the individual reactions of the CAC are under independent control.

Availability of ADP

If ADP is limited so is the rate of ADP phosphorylation. This in turn, limits the rate of electron transfer from NADH to oxygen so not enough NAD^+ is available for the citric acid cycle. Remember that NAD^+ is a substrate for three of the CAC enzymes (isocitrate dehydrogenase, α-ketoglutarate dehydrogenase, and malate dehydrogenase). Thus an inadequate rate of ADP entry into the mitochondrion will mean that the concentration of NAD^+ is too low to stimulate the CAC. In other words, the CAC is tightly linked to the rate of electron transfer to oxygen and the rate of phosphorylation of ADP.

Regulation of Irreversible CAC Enzymes

Three of the CAC enzymes catalyze irreversible reactions and, as is typical in metabolism, are regulated. This regulation is useful because, even if the overall CAC is operating at a low rate, the irreversible reactions could continue to use up their substrates if they are available. Regulation can occur by allosteric mechanisms or by substrate availability. In the case of the two dehydrogenases (isocitrate dehydrogenase and α-ketoglutarate dehydrogenase), an increase in the [NADH]/[NAD^+] ratio means that there is a reduced amount of the substrate NAD^+ because the total amount of the nicotinamide adenine dinucleotide coenzymes would be constant. Accordingly, the activities of isocitrate and α-ketoglutarate dehydrogenase would be reduced due to a limiting supply of one substrate.

Citrate synthase (CS) should be controlled because otherwise it could consume the available acetyl CoA and oxaloacetate. To prevent this, NADH is a negative allosteric inhibitor of CS. NADH binds to an allosteric site on CS, and in so doing, increases the K_m of CS for its substrate, acetyl CoA. Also, citrate is a competitive inhibitor for oxaloacetate at the active site of CS. Therefore, as citrate concentration increases, it blocks access of oxaloacetate at the active site of CS.

Isocitrate dehydrogenase (ICDH) is inhibited by NADH at a negative allosteric site. Thus, at rest, when NADH concentration is high, ICDH is inhibited (just like CS). In addition, ICDH is activated by calcium ions. In active muscle, cytoplasmic Ca^{2+} increases, and some enters the matrix where it activates ICDH. This process reflects a useful connection between the cytosol and the mitochondrial matrix. The calcium enters the mitochondrion by moving through a uniporter, a special protein that allows a substance to cross a membrane down its concentration gradient. A sodium-calcium antiport is used to remove the calcium from the matrix—one Ca^{2+} out for two Na^+ in.

The α-ketoglutarate dehydrogenase is inhibited by NADH (just like CS and ICDH). In addition, succinyl CoA is a competitive inhibitor for CoA. Thus this enzyme cannot tie up the available CoA in the matrix. In addition, like ICDH, Ca^{2+} activates α-ketoglutarate dehydrogenase.

Several terms are often used by physiologists. The mitochondrial redox state reflects the relative concentrations of NADH and NAD^+, which may be expressed as the concentration ratios [NADH]/[NAD^+], sometimes written [NAD^+]/[NADH]. The redox state of the mitochondrion is much more reduced than that of the cytoplasm. Thus relatively more NADH exists in the matrix than in the cytoplasm.

The term reducing equivalents means a pair of electrons that can be transferred to oxygen, for example, NADH or $FADH_2$.

In many athletic or recreational activities, glycolysis is an important source of ATP, generating the product lactate. However, all the carbon atoms in lactate will eventually appear as CO_2 because oxidative phosphorylation is the final pathway for all fuels, even though there may be a temporary production of lactate during vigorous activity. The carbon atoms of lactate can also be used to make fat or glucose and their final route will be through the citric acid cycle where electrons are transferred to oxygen to make water.

Regulation of Pyruvate Oxidation

As mentioned previously, pyruvate—formed in glycolysis in the cytosol—can enter the mitochondrion where it is converted to acetyl CoA. Pyruvate crosses the inner membrane via a symport in which a hydrogen ion (proton) goes the same direction, down its gradient. The reaction for pyruvate oxidation in the matrix is as follows:

$$\text{pyruvate} + NAD^+ + \text{CoA} \xrightarrow{\text{pyruvate dehydrogenase}} \text{acetyl CoA} + NADH + H^+ + CO_2$$

This reaction is catalyzed by a matrix enzyme complex known as pyruvate dehydrogenase, abbreviated PDH. The PDH complex is composed of five types of enzyme subunits and three tightly bound coenzymes. The coenzymes are lipoic acid (synthesized in the body), thiamine pyrophosphate (TPP), and FAD. Two other coenzymes can also be seen in this reaction, CoA and NAD^+. As mentioned previously, the PDH reaction is similar to that of α-ketoglutarate dehydrogenase in that both use the same five coenzymes, although the subunits of the enzyme complexes are different.

The PDH reaction must be carefully regulated because the irreversible conversion of pyruvate to acetyl CoA means that a potential gluconeogenic precursor is lost, that is, pyruvate can be converted to glucose in the liver, but acetyl CoA cannot. Because the brain needs glucose and is treated biochemically as the most important tissue in the body, the PDH reaction must be regulated to spare pyruvate from being irreversibly lost. Regulation of PDH occurs via phosphorylation and dephosphorylation (see Figure 9.7). To prevent the unnecessary oxidation of pyruvate to acetyl CoA when other fuels such as fatty acids and ketone bodies can provide the acetyl CoA, PDH is phosphorylated by PDH kinase into the inactive form, PDH-P. The inactivation of PDH occurs during starvation, fasting, or low intensity exercise.

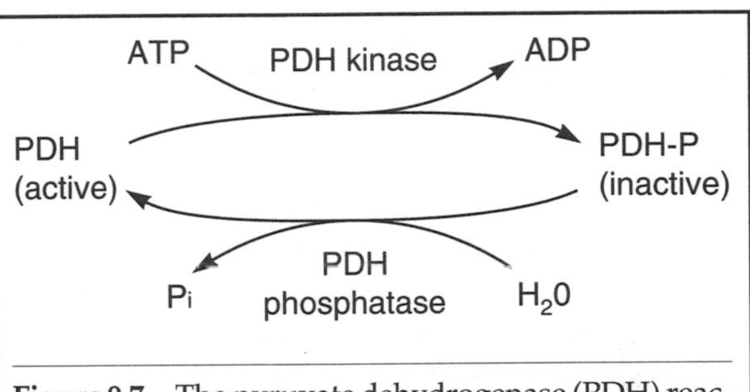

Figure 9.7 The pyruvate dehydrogenase (PDH) reaction is regulated by reversible phosphorylation and dephosphorylation of one subunit. In the dephosphorylated state, PDH is active. When a phosphate group is transferred from ATP to PDH, it becomes inactive. Phosphorylation is catalyzed by PDH kinase, whereas the phosphate group is removed by PDH phosphatase.

PDH kinase is one component of the PDH complex. It is activated when an excess amount of acetyl CoA and NADH are in the matrix. Under these conditions high intramitochondrial ratios of acetyl CoA/CoA and NADH/NAD^+ exist. These increased ratios reflect an abundance of substrate for the citric acid cycle and electron transport chain, respectively. PDH kinase is also inactivated by high intramitochondrial Ca^{2+} concentration, which results from a sustained increase in cytosolic calcium ions, for example, by continuous contractile activity by a muscle cell.

PDH phosphatase is activated by an increase in intramitochondrial Ca^{2+} concentration. Thus, PDH kinase is inactivated and PDH phosphatase is activated by calcium. As a result, for muscle the state of PDH activity is affected both by the metabolic state of the muscle fiber and the contractile state of the fiber.

Summary

The citric acid cycle (Krebs cycle or tricarboxylic acid cycle) is the pathway that removes the last carbon atoms in the form of CO_2 from all the body's fuels while electrons associated with the hydrogen atoms of these fuels are used to reduce the coenzymes NAD^+ and FAD. Because the electron transfer chain (respiratory chain) is the pathway that removes electrons from the reduced coenzymes NADH and $FADH_2$ and transfers them to oxygen, the citric acid cycle and electron transfer chain are tightly related. The citric acid cycle is a circular pathway catalyzed by eight enzymes, all of which are located in the mitochondrial matrix except succinate dehydrogenase. Acetyl groups containing two carbon atoms enter the pathway, attached to CoA, and attach to oxaloacetate to form citrate. Because one turn of the cycle yields two CO_2, the citric acid cycle neither produces nor consumes oxaloacetate or any other intermediate of the cycle. The citric acid cycle nets 12 ATP molecules because it produces three NADH, one $FADH_2$, and one GTP.

Oxidation of acetyl groups in the citric acid cycle and oxidation of reduced coenzymes in the electron transfer chain are tightly coupled to the phosphorylation of ADP to make ATP. Therefore, regulation of the citric acid cycle and the electron transfer chain is closely linked to the rate of ATP utilization. The rate of the citric acid cycle is also controlled by the three enzymes catalyzing irreversible steps, citrate synthase, isocitrate dehydrogenase, and alpha-ketoglutarate dehydrogenase, as well as by the rate at which the reduced coenzyme products of the cycle, NADH and $FADH_2$, are reoxidized in the respiratory chain. For skeletal muscle, the rate ADP is produced by ATPase enzymes dictates the overall rate of oxidative phosphorylation. In addition, oxygen availability can limit the overall rate of fuel oxidation because oxygen is the final electron acceptor. The rate of NADH formation in the citric acid cycle can limit the overall rate of oxidative phosphorylation in skeletal muscle if carbohydrate available to exercising muscle is limited, forcing the muscle to use fat as the exclusive fuel. Pyruvate, generated by glycolysis, can be decarboxylated in mitochondria to acetyl CoA. However, this step is irreversible and removes a three-carbon molecule that is a potential source of glucose. Therefore, the pyruvate dehydrogenase reaction is carefully regulated by phosphorylation, which inhibits the enzyme, and dephosphorylation, which activates it.

CHAPTER 10

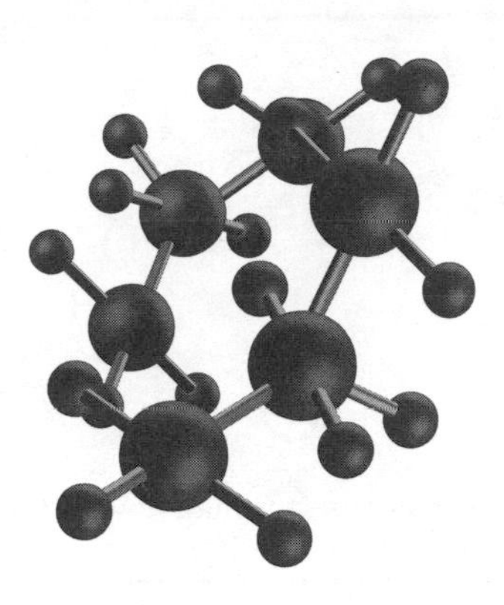

Lipid Metabolism

Most people are well aware that fats (or lipids) are very important fuels in the body, but lipids are also important structural elements. To understand the importance of lipids one must learn about the various kinds of lipids and their metabolism.

Types of Lipids

Lipids are categorized as fatty acids, triglycerides, phospholipids, and sterols, whose role in the body is largely determined by their chemical structure.

Fatty Acids

Fatty acids are long-chain carboxylic acids, usually known by their trivial names. They usually have an even number of carbon atoms and can be saturated, with no carbon-to-carbon double bonds, or unsaturated, with one or more carbon-to-carbon double bonds. Figure 10.1 gives examples of saturated and unsaturated fatty acids. Palmitic and stearic acids are saturated fatty acids. A shorthand notation for describing their structure appears in the square brackets. The first number reveals the number of carbon atoms and the second indicates the number of double bonds. Oleic and linoleic acids are unsaturated fatty acids. The superscript numbers with the Greek symbol Δ (delta) indicate the position of the double bonds, beginning with the carboxyl group as carbon one. The configuration of the double bonds in naturally occurring fatty acids is cis, as is shown in Figure 10.1.

Linoleic acid is an essential fatty acid that we must get in the diet. It is also known as a PUFA, that is, a polyunsaturated fatty acid. Linolenic acid is [18:3($\Delta^{9,12,15}$)], and is called an omega 3 or n-3 fatty acid, that is, the last double bond begins three carbons from the end carbon, or carbon 18. Note that linoleic acid is an omega six or n-6 fatty acid. The n-3 (or omega three) fatty acids are reputed to offer special protection to people by lowering blood lipid concentration. Linolenic acid is also considered a dietary essential fatty acid.

Triglycerides

Although they should be called triacylglycerols, the term triglyceride is more commonly used. An older term, neutral fat, is also used. Figure 10.2 shows these molecules as triesters, made from the combination of the trialcohol glycerol and three fatty acids. In the body, stored triglycerides rarely have the same fatty acid attached at all three positions on the glycerol. Stored fats have both saturated and unsaturated fatty acids.

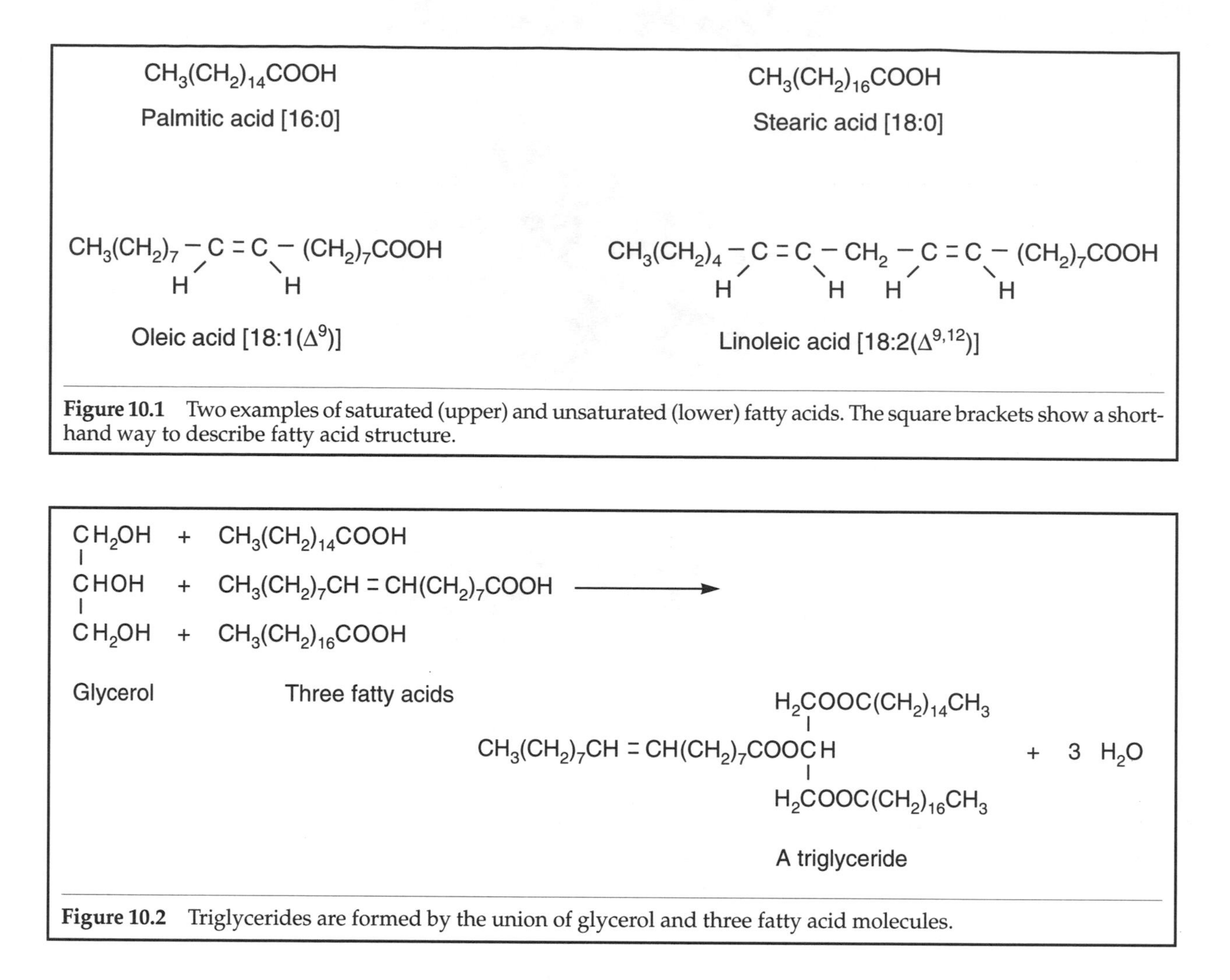

Figure 10.1 Two examples of saturated (upper) and unsaturated (lower) fatty acids. The square brackets show a short-hand way to describe fatty acid structure.

Figure 10.2 Triglycerides are formed by the union of glycerol and three fatty acid molecules.

The physical state of triglycerides depends on the length of the carbon chain in the fatty acids and the number of double bonds. Shorter length fatty acids as well as more double bonds lower the melting point. Thus, vegetable fats with polyunsaturated fatty acids are liquids, whereas the fat on the side of a steak, which contains more saturated fatty acids, is a solid. Palm oil is a liquid, despite having mainly saturated fatty acids, because the fatty acids have 10 to 12 carbon atoms as opposed to 16 to 18 in most other triglycerides. Triglycerides and the fatty acids found in triglycerides are insoluble in water due to the large degree of hydrophobic hydrocarbon components. The hydrophobic nature of triglycerides makes them ideal for storing energy because by weight they have more chemical potential energy than other fuel molecules, such as carbohydrates or protein.

Diglycerides or, more correctly, diacylglycerols have only two fatty acids attached to the glycerol. Monoglycerides or monoacylglycerols have only one.

Phospholipids

Many phospholipids are derivatives of phosphatidic acid, whose structure is illustrated in Figure 10.3. Different groups can be attached to the phosphate in phosphatidic acid. If choline is attached, the molecule is called phosphatidyl choline or, commonly, lecithin (see Figure 10.3). Phospholipids have a hydrophilic part and a hydrophobic part. The hydrophilic part is due to polar chemical bonds and charged groups on the phospholipid. The hydrophobic part is the long hydrocarbon tail of the fatty acids, which can contain more than 16 carbon atoms.

Phosphatidyl inositols are found in membranes, where they play an important role in cellular regulation. Figure 10.4 shows the structure of inositol. When phosphorylated at carbons four and five and attached to phosphatidic acid, it is known as phosphatidyl inositol-4,5-bisphosphate, abbreviated PtdIns(4,5)P_2. Hydrolysis of PtdIns(4,5)P_2 produces inositol 1,4,5-trisphosphate, abbreviated Ins(1,4,5)P_3, and a diacylglycerol (DG).

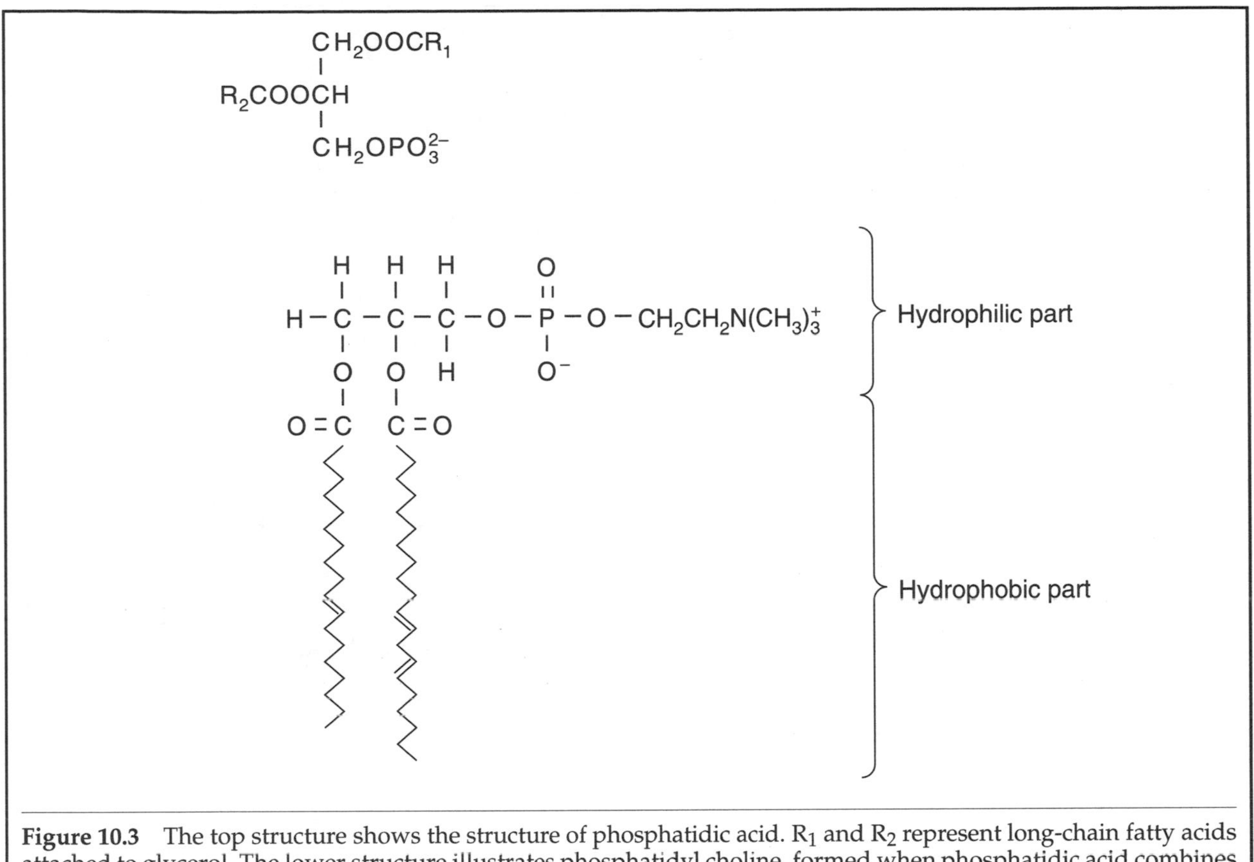

Figure 10.3 The top structure shows the structure of phosphatidic acid. R_1 and R_2 represent long-chain fatty acids attached to glycerol. The lower structure illustrates phosphatidyl choline, formed when phosphatidic acid combines with choline. The structure has been rotated to illustrate a hydrophilic part with polar bonds and charged groups and a long hydrophobic part, due to the tails of the fatty acids, shown in chemical shorthand.

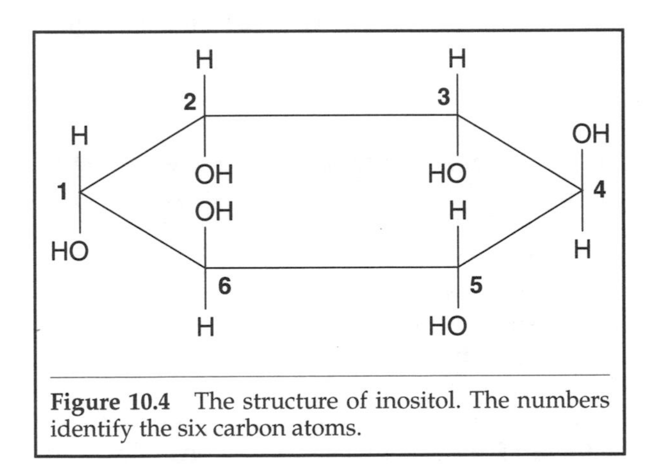

Figure 10.4 The structure of inositol. The numbers identify the six carbon atoms.

PtdIns(4,5)P_2 and Ins(1,4,5)P_3 are important in signal transduction. Certain intercellular signals, such as hormones or nerve impulses, can activate a cell membrane enzyme known as phospholipase C. This enzyme can then hydrolyze cell membrane phospholipids (PtdIns[4,5]P_2) to produce Ins(1,4,5)P_3 and a diacylglycerol. The Ins(1,4,5)P_3 can act on the endoplasmic reticulum of the cell to release calcium ions, which can then activate other enzymes. The diacylglycerol can also activate an enzyme known as protein kinase C, which in turn can phosphorylate other enzymes. In this way, a neural signal or hormone can cause specific changes in a cell. We call this process signal transduction because the nerve impulse or hormone (called the first messenger) acts at the cell membrane by binding to specific receptors, resulting in the formation of second messengers inside the cell. Second messengers, such as Ins(1,4,5)P_3, diacylglycerol, and Ca^{2+}, can promote the phosphorylation of certain proteins, which stimulate or inhibit their functions, generating specific responses within the cell. Malfunctions in the signal transduction process underlie many harmful diseases in the body. Figure 10.5 provides an overview of the phosphatidyl inositol signaling pathway.

Fat Stores

Fat is stored as triacylglycerol (triglyceride) in fat cells. It is considered a long-term energy store in contrast to glycogen, which is considered a short-term energy store. A normal adult can store about 2,000 to 3,000

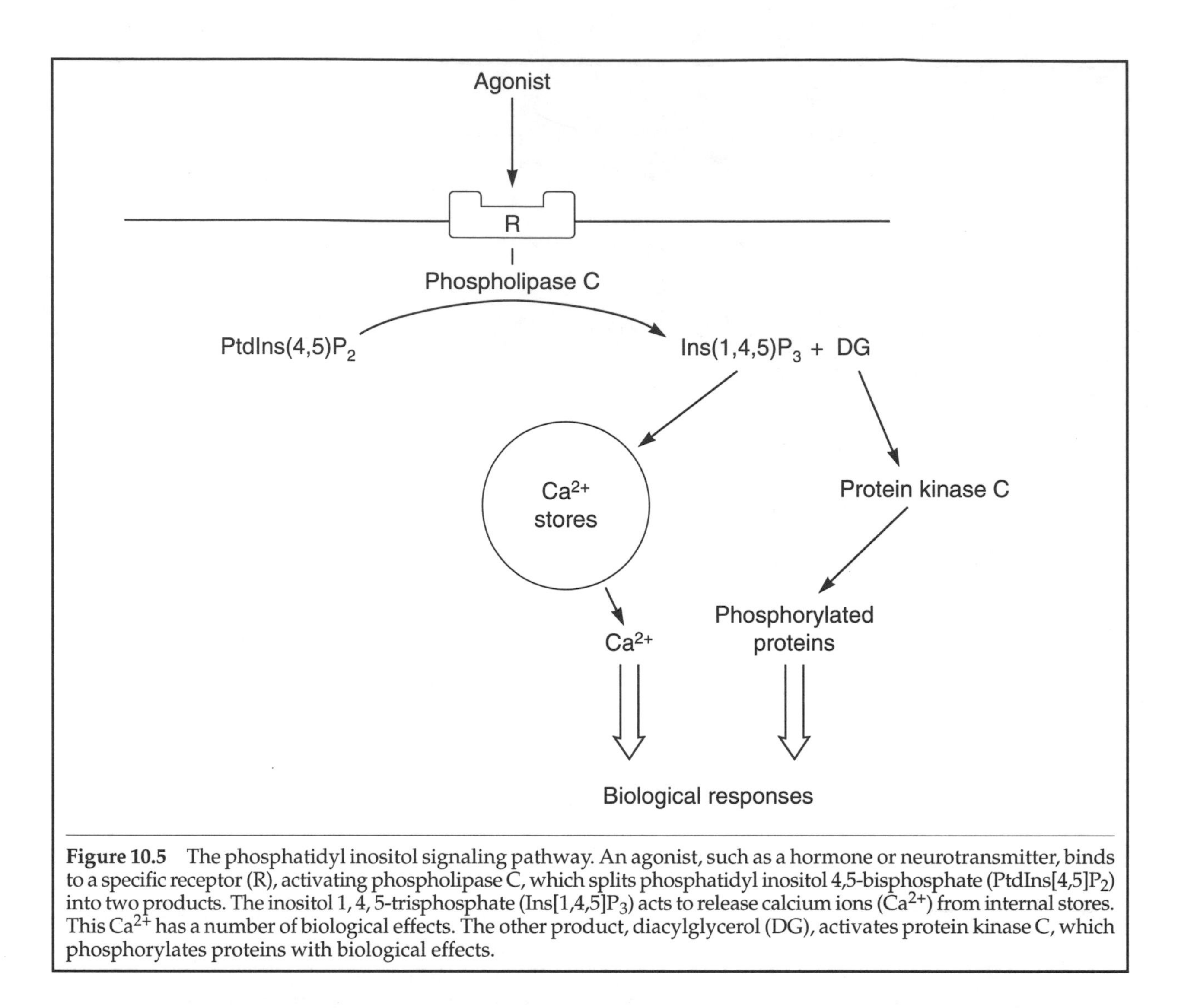

Figure 10.5 The phosphatidyl inositol signaling pathway. An agonist, such as a hormone or neurotransmitter, binds to a specific receptor (R), activating phospholipase C, which splits phosphatidyl inositol 4,5-bisphosphate (PtdIns[4,5]P_2) into two products. The inositol 1, 4, 5-trisphosphate (Ins[1,4,5]P_3) acts to release calcium ions (Ca^{2+}) from internal stores. This Ca^{2+} has a number of biological effects. The other product, diacylglycerol (DG), activates protein kinase C, which phosphorylates proteins with biological effects.

kilocalories (kcal) of glycogen in liver and muscle, but even a lean male or female has more than 75,000 kcal stored as triglyceride in fat cells. Stored energy or the energy content of food is properly expressed in units of kilojoules (kJ), but use of the older kcal is still common. One kcal is equivalent to 4.2 kJ. The major storage form of triglycerides is a liquid droplet, occupying much of the volume of the fat cell.

Formation of Triacylglycerols

Formation of triacylglycerols is a simple process, illustrated as a brief sequence of reactions in Figure 10.6. This process is also known as esterification because it forms the triester, triglyceride. The fatty acid substrates are attached to CoA and are known as fatty acyl CoAs. This process is favored after eating a meal, when the body receives fatty acids and glucose from the food. The initial precursor, glycerol 3-phosphate, is derived from dihydroxyacetone phosphate (discussed under the section on glycolysis). Glucose from blood is taken up by fat cells and, through partial breakdown via the glycolytic reactions, is converted to dihydroxyacetone phosphate. Reduction of dihydroxacetone phosphate generates glycerol 3-P. Also following a meal, the insulin concentration in the blood will increase, favoring entry of glucose into fat cells. The 1-lysophosphatidic acid is simply phosphatidic acid without a fatty acid attached to carbon 2 of glycerol.

Mobilization of Fat

Breaking down triglycerides to yield their energy-rich fatty acids and glycerol is known as triglyceride hydrolysis or lipolysis. This process is favored under conditions of exercise, low calorie dieting, fasting or starvation, and when the body is cold. Lipolysis involves three hydrolysis reactions, each catalyzed by hormone-sensitive lipase (see Figure 10.7). The last

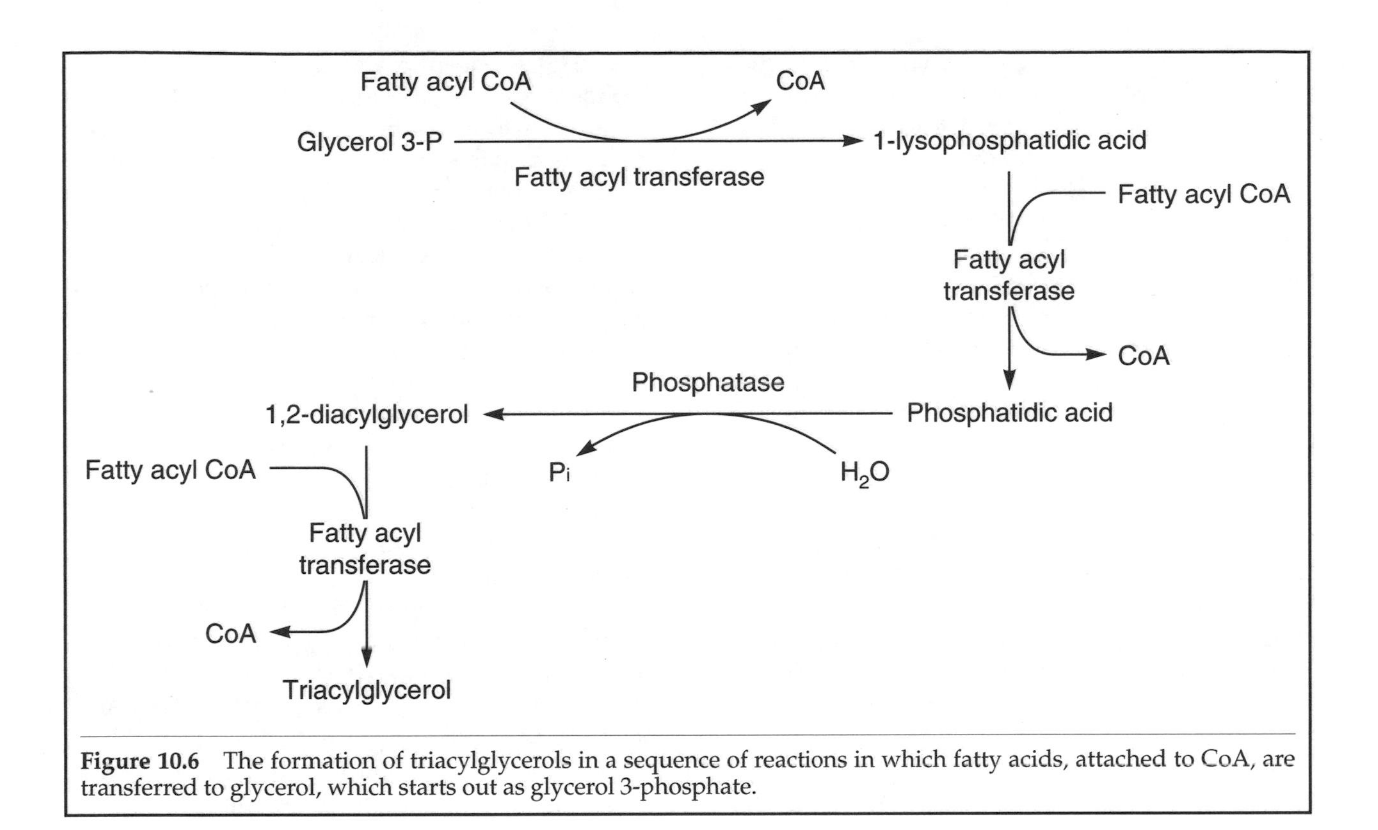

Figure 10.6 The formation of triacylglycerols in a sequence of reactions in which fatty acids, attached to CoA, are transferred to glycerol, which starts out as glycerol 3-phosphate.

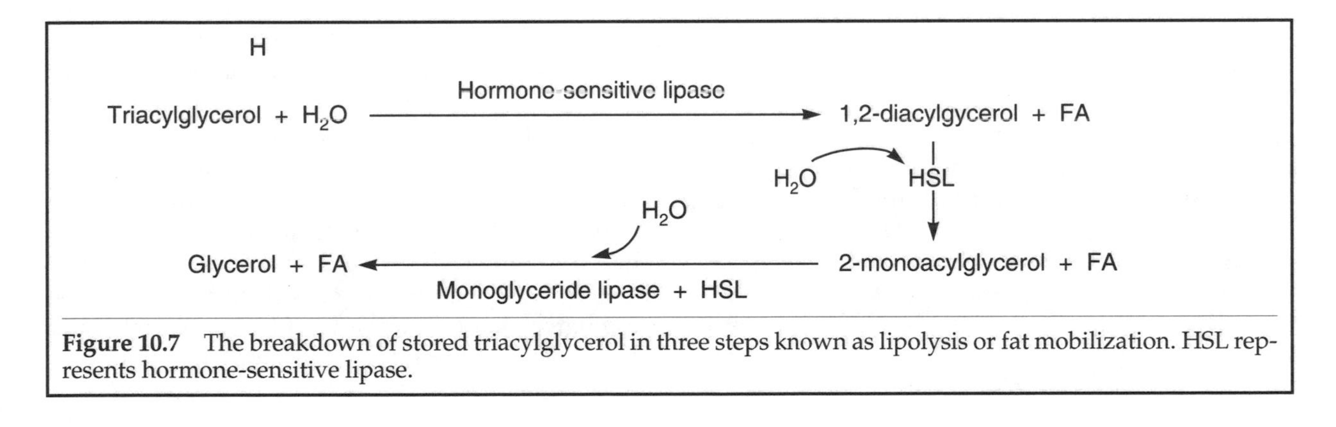

Figure 10.7 The breakdown of stored triacylglycerol in three steps known as lipolysis or fat mobilization. HSL represents hormone-sensitive lipase.

hydrolysis reaction is also assisted by a monoglyceride lipase. The numbers refer to the carbon atoms of glycerol. For example, a 2-monoglyceride (2-monoacylglycerol) has a single fatty acid attached to the glycerol molecule at the middle carbon atom.

Both lipolysis and formation of triacylglycerols take place in the cytosol of fat cells or adipocytes. Fatty acids released when triacylglycerol molecules are hydrolyzed have two major fates: (a) reesterification to a triglyceride, following conversion to a fatty acyl CoA; or (b) exit from the fat cell.

When fatty acids exit a fat cell and enter the blood, they attach to the blood protein albumin. Fatty acids attached to albumin are known as free fatty acids or FFA. Remember, fatty acids are long molecules, mainly 16 to 18 carbon atoms in length, that are not soluble in the aqueous medium of the blood. Thus they must attach to albumin. The albumin-bound FFA circulate in the blood and enter other cells, where they are used as energy substrates.

Glycerol, released when a triacylglycerol is hydrolyzed, cannot be reused by the fat cell. Instead, the glycerol leaves and circulates in the blood. Because glycerol is not reused by the fat cell, whereas fatty acids may be reesterified, release of glycerol is taken as the index of lipolysis. Glycerol has three OH groups, making it is quite soluble in the blood. It is taken up by tissues that contain glycerol kinase, which phosphorylates glycerol to glycerol 3-phosphate. The major tissue where this occurs is the liver. Indeed, when glycerol is taken up by liver cells and phosphorylated to glycerol 3-P, it can then be used to form glucose by the process of gluconeogenesis, which we will discuss later.

Although we will look only at lipolysis in fat cells, this process takes place in other areas of the body. For example, hydrolysis (digestion) of dietary triglyceride occurs in the small intestine, catalyzed by pancreatic lipase. Hydrolysis of triglyceride in blood lipoproteins is catalyzed by lipoprotein lipase, an enzyme located on the walls of capillaries. The fatty acids released by lipoprotein lipase can be taken up by adjacent cells where they will be stored as triglyceride, an important source of fatty acids for fat cells.

A constant turnover of triglyceride exists in fat calls with triglyceride formation and lipolysis. Which process predominates depends on a number of factors. For example, following a meal, formation of triglycerides is dominant because the meal provides precursors (glucose and fatty acids), and the insulin level is high in the blood. During exercise, lipolysis is favored. Regulatory mechanisms operate to control whether lipolysis or esterification dominate because they take place in the same compartment of the cell.

Not much

Regulation of Triglyceride Turnover

The human body can store a large amount of triacylglycerol in fat and muscle cells. Because triacylglycerol is an important fuel for skeletal and heart muscle and because a limited supply of carbohydrate is stored in the body, fat must be utilized whenever its oxidation can produce the ATP needed. Regulatory systems operate to promote fat usage whenever possible and to spare carbohydrate. During exercise such regulation is extremely important because the rate of ATP hydrolysis during vigorous muscle activity (and thus the need for fuel) is so much greater than during normal daily activities.

Lipid metabolism must be controlled in the fat cell to prevent what we call futile cycling, for it is energetically wasteful to have both lipolysis and esterification operating maximally at the same time. The net effect would be hydrolysis of ATP because it is used to make glycerol 3-phosphate from glucose and is needed to attach fatty acids to CoA. Instead, regulatory mechanisms ensure that one process is dominant and the other depressed. For example, during exercise, esterification (triglyceride formation) is inhibited, whereas lipolysis is accelerated. Following a meal, lipolysis is inhibited, whereas esterification is accelerated. Regulation takes place primarily at the level of hormone-sensitive lipase (HSL), controlled by phosphorylation-dephosphorylation. When HSL is phosphorylated, it is active; when it is dephosphorylated, it is inactive.

Figure 10.8 shows the mechanism for regulating metabolism in the fat cell. Hormone-sensitive lipase (HSL) is activated by accepting a phosphate group from ATP in a reaction performed by the catalytic subunit of cyclic AMP (cAMP)-dependent protein kinase, identified as C. The cAMP-dependent protein kinase is also known as protein kinase A or A kinase. HSL is inactivated by removal of the phosphate group, catalyzed by a protein phosphatase. Normally, protein kinase A exists with two types of subunits. The catalytic subunits (C) are inactive when they are bound to the regulatory subunits (R) by noncovalent bonds. If the R subunits bind cAMP, they dissociate from the C subunits, and the latter become active.

Cyclic AMP is formed when the enzyme adenylate cyclase (AC) is activated. When activated, adenylate cyclase converts an ATP molecule into 3′-5′ cAMP and inorganic pyrophosphate (PPi). The cAMP has a single phosphate group that is attached both to the 5′ and the 3′ carbon atom of the sugar ribose, thus the term cyclic. The hormone epinephrine (also known as adrenaline) and the sympathetic nervous system neurotransmitter norepinephrine (also known as noradrenaline) can bind to a beta-receptor (R_β). Then, by interacting with a stimulatory G protein (G_s), adenylate cyclase is activated. Another type of receptor, the alpha-2 receptor ($R_{\alpha 2}$) can bind certain agonists that then interact with an inhibitory G protein (G_i) to decrease adenylate cyclase activity. When either epinephrine or norepinephrine bind to the beta receptor, AC is active, converting ATP to cAMP and PPi. As mentioned, cAMP then binds to the regulatory subunit of protein kinase A, releasing the catalytic subunit (C). This C subunit then phosphorylates HSL to make it active. The balance between the number of stimulatory (R_β) and inhibitory ($R_{\alpha 2}$) receptors helps determine the final response of fat cells to lipolytic effectors.

The PPi formed when ATP is converted into cAMP is hydrolyzed by inorganic pyrophosphatase to form two Pi. Furthermore, cAMP is not allowed to linger in the cell. An enzyme known as cyclic AMP phosphodiesterase (PDE) breaks the bond that attaches the phosphate to the 3′ carbon of ribose, producing 5′-AMP, thus stopping the cAMP signal. Cyclic AMP is also known as a second messenger, the first messenger being the hormone epinephrine or the neurotransmitter norepinephrine. Both first messengers act at the cell membrane by binding to their specific receptors, resulting in the formation of a second messenger (cAMP in this case), which then

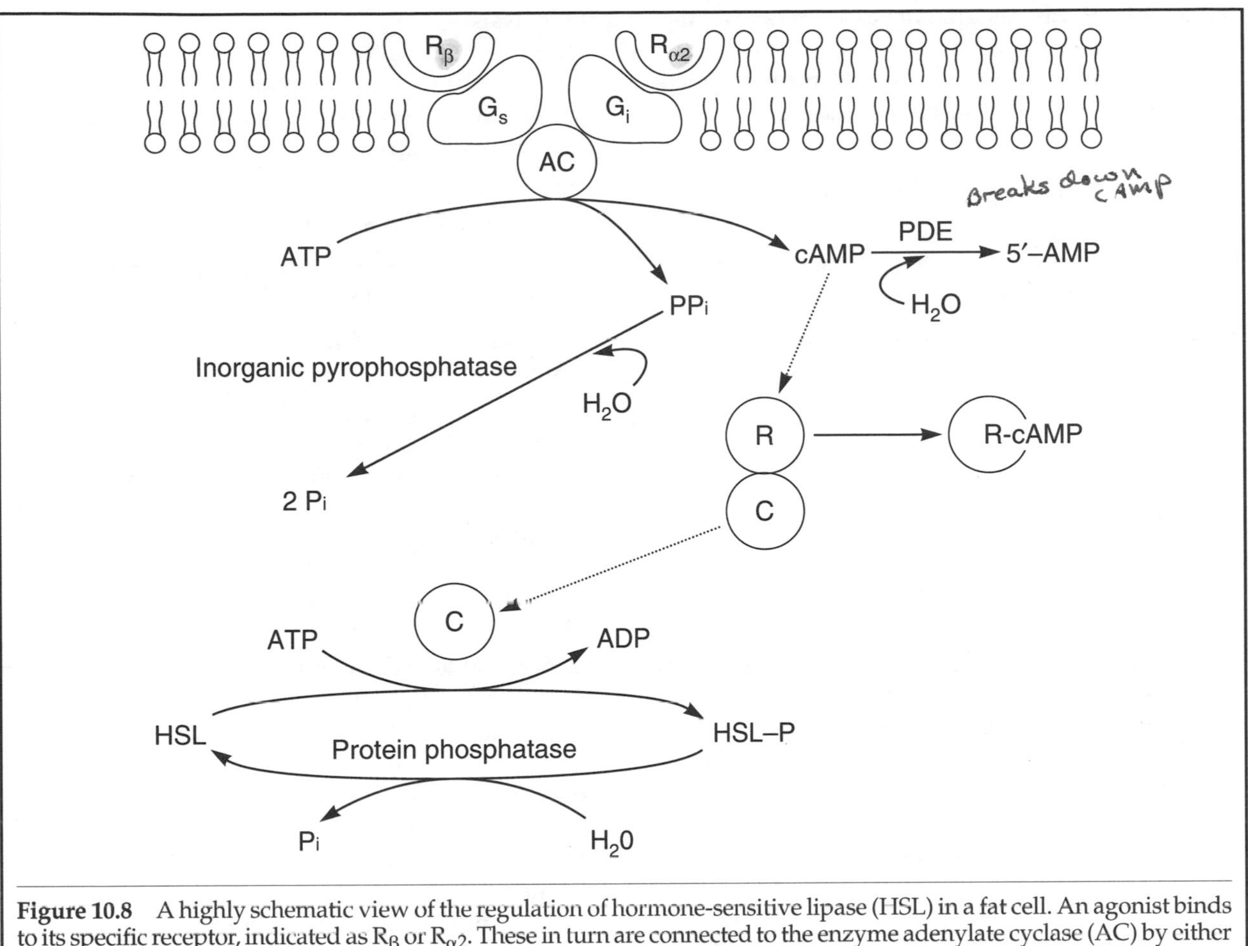

Figure 10.8 A highly schematic view of the regulation of hormone-sensitive lipase (HSL) in a fat cell. An agonist binds to its specific receptor, indicated as R_β or $R_{\alpha 2}$. These in turn are connected to the enzyme adenylate cyclase (AC) by either a stimulatory (G_s) or inhibitory (G_i) G protein, respectively. Receptors, G protein, and AC are found in the cell membrane, shown as the lipid bilayer. Agonists binding to R_β activate AC. Cyclic AMP (cAMP), formed by AC, activates a protein kinase consisting of regulatory (R) and catalytic (C) subunits. The catalytic subunit activates the inactive form of HSL by attaching a phosphate group, donated by ATP. The active or HSL-P form now breaks down triacylglycerol to fatty acids and glycerol. The cAMP signal ceases when cAMP phosphodiesterase (PDE) breaks it down to 5′-AMP.

stimulates some cellular process (lipolysis in this example). Referring back to Figure 10.5 reveals that agonist binding to other receptors can generate other second messengers, such as inositol 1,4,5 trisphosphate, diacylglycerol, and calcium ions.

Table 10.1 lists hormones that increase lipolysis, decrease lipolysis, or stimulate triglyceride formation. A group of compounds known as methyl xanthines also have lipolytic effects. Caffeine, the best known methyl xanthine, inhibits phosphodiesterase activity, thus strengthening and prolonging the effect of cAMP. Endurance athletes often ingest caffeine before competition to increase the blood concentration of fatty acids, thereby hoping to reduce their initial reliance on carbohydrate as a fuel and prolong the time they can perform at a high level.

In addition to decreasing the rate of lipolysis, insulin also promotes triacylglycerol synthesis by promoting glucose uptake into fat cells. Glucose is the precursor for making glycerol 3-phosphate, and insulin probably increases the activity of the fatty acyl transferase enzyme (see Figure 10.6) to increase transfer of fatty acids to glycerol 3-phosphate.

Intracellular Transport of Fatty Acids

Fatty acids travel in the blood attached to albumin; in this form they are known as free fatty acids (FFA). Fatty acids either enter cells with the aid of a transporter or simply diffuse across the cell membrane. To get the energy available in the fatty acid molecule (more than nine kcal/g versus only four kcal/g for carbohydrates), they must be transferred from the cytoplasm into the mitochondrion. Because of their insolubility, fatty acids in the cell cytoplasm of many

Table 10.1 Hormones That Increase or Decrease the Rate of Lipolysis

Hormone	Source	Effect
Epinephrine	Adrenal medulla	Increases lipolysis when bound to β-receptors. Decreases lipolysis when bound to α-2 receptors.
Norepinephrine	Sympathetic nervous system	Increases lipolysis when bound to β-receptors. Decreases lipolysis when bound to α-2 receptors.
Thyroid hormones	Thyroid gland	Increase lipolysis by increasing the number of β-receptors on fat cells.
Growth hormone	Pituitary gland	Increases the rate of lipolysis; mechanism is not known for sure.
Insulin	Pancreas	Decreases lipolysis by decreasing the number of β-receptors on the surface of the fat cell, increasing the activity of phosphodiesterase to lower cAMP concentration, and promoting fatty acid incorporation into triglyceride.

tissues are attached to a fatty acid binding protein (FABP). In muscle, fatty acids attached to FABP can be those absorbed from the blood or those released when muscle triacylglycerol is hydrolyzed.

* Formation of Acyl CoA

Before a fatty acid can enter the mitochondrion for oxidation, it must first have a CoA attached to it. An energy-rich fatty acyl CoA is formed using the free energy of hydrolysis of ATP:

$$\text{fatty acid} + \text{ATP} + \text{CoA} \xrightarrow[\text{Mg}^{2+}]{\text{acyl CoA synthetase}} \text{fatty acyl CoA} + \text{AMP} + \text{PPi}$$

This reaction is essentially irreversible because the PPi is hydrolyzed by inorganic pyrophosphatase to two Pi (not shown), which drives the reaction to the right. The fatty acyl CoA that is formed is an energy-rich molecule, much like acetyl CoA. Formation of a fatty acyl CoA costs two ATP because two phosphates are removed from ATP. Because the pK_a for fatty acids is much less than the cytosolic pH, the fatty acids exist as anions.

*Transport as Acyl Carnitine

Formation of fatty acyl CoA (often called simply acyl CoA) occurs in the cytosol, whereas oxidation of the fatty acyl CoA occurs in the mitochondrial matrix. However, the mitochondrial inner membrane is impermeable to CoA and its derivatives, which permits separate regulation of CoA compounds in mitochondrial and cytosolic compartments. Transport of fatty acyl CoA into the mitochondrial matrix occurs using three different proteins and the small molecule carnitine (see Figure 10.9). Only fatty acids attached to carnitine are able to cross the inner mitochondrial membrane.

Carnitine, formed from the amino acids lysine and methionine, can cross the inner membrane, and a fatty acyl form of carnitine can also cross in the opposite direction. The carnitine acyl carnitine translocase is called the carnitine acyl carnitine antiport because this membrane protein transfers two different substances across the inner membrane in opposite directions. The enzyme carnitine acyl transferase, located both on the cytosolic and matrix side of the inner membrane, transfers the fatty acyl group from CoA to carnitine and from carnitine to CoA, respectively. Carnitine deficiency, due to the body's inability to convert lysine into carnitine, is not an unusual metabolic disease. Patients have muscle weakness and poor exercise tolerance due to accumulation of triglyceride in muscle and the inability to oxidize fatty acids. Carnitine supplementation by some athletes has also become a fad. Supplemental carnitine is supposed to enhance lipid oxidation and spare carbohydrate utilization, and part of the rationale for using it is that exercise causes an increased urinary excretion of carnitine, possibly leading to reduced levels in active muscles. However, reliable experimental data supporting the use of supplementary carnitine are difficult to find.

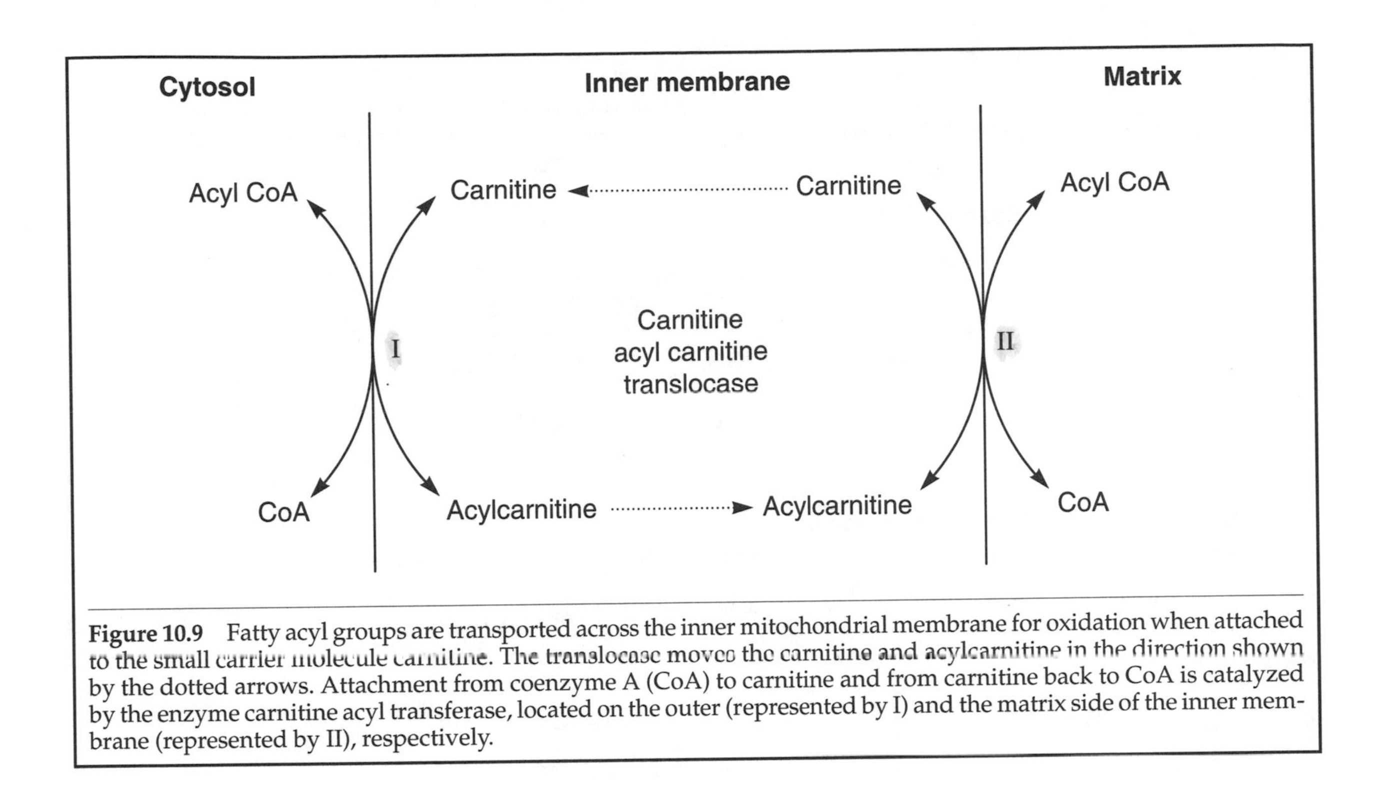

Figure 10.9 Fatty acyl groups are transported across the inner mitochondrial membrane for oxidation when attached to the small carrier molecule carnitine. The translocase moves the carnitine and acylcarnitine in the direction shown by the dotted arrows. Attachment from coenzyme A (CoA) to carnitine and from carnitine back to CoA is catalyzed by the enzyme carnitine acyl transferase, located on the outer (represented by I) and the matrix side of the inner membrane (represented by II), respectively.

Oxidation of Saturated Fatty Acids

The initial process in the oxidation of fatty acids is known as β oxidation, which occurs in the matrix of the mitochondrion. Beta oxidation begins as soon as the fatty acyl CoA appears in the matrix, using repeated cycles of four steps, so that with each cycle, the fatty acyl CoA is broken down to form a new fatty acyl CoA, shortened by two carbon atoms, plus an acetyl CoA. Figure 10.10 shows the four steps in β oxidation.

The first reaction is catalyzed by the enzyme acyl CoA dehydrogenase, located on the inner membrane of the mitochondrion. The asterisk (*) in Figure 10.10 indicates the β-carbon atom of the fatty acyl CoA, also carbon three, numbering from the acyl carbon or carbon one of the fatty acid. In this reaction, two hydrogen atoms, each with an electron, are removed and transferred to FAD, making it $FADH_2$. These electrons are then passed to coenzyme Q. The removal of the two hydrogen atoms results in the formation of a C-to-C double bond between the α-C (or carbon two) and the β-C (or carbon three) of the fatty acyl group. The double bond is trans because the two hydrogens are on opposite sides.

In the second reaction trans-enoyl CoA is hydrated, accepting a water molecule. The OH part of the water is added to the β-C while the other hydrogen atom of water is added to the α-C. The product is a 3-hydroxyacyl CoA. The enzyme, enoyl CoA hydratase, is located in the matrix. In the third reaction, catalyzed by the matrix enzyme 3-hydroxyacyl CoA dehydrogenase, the β-C of the hydroxyacyl CoA is oxidized, losing a hydride ion and a proton. Such oxidation reactions always use the nicotinamide coenzymes, NAD^+ in this case. This reaction changes the β-carbon, or carbon three, to a ketone and is β oxidation.

In the last step CoA attaches to the β-carbon (or C 3), which allows carbons one and two to come off as an acetyl CoA, leaving a new acyl CoA shortened by two carbon atoms. The enzyme responsible for the fourth reaction is acyl CoA thiolase, called thiolase because the CoA contains a terminal SH (thiol) group, as discussed in the beginning of the citric acid cycle.

The new acyl CoA now undergoes the same four enzyme-catalyzed steps, creating another acetyl CoA and a new acyl CoA, again shortened by two carbon atoms. Eventually the original fatty acyl CoA, which in Figure 10.10 contained 18 carbon atoms, is reduced to nine acetyl CoA by eight cycles of β oxidation. The acetyl CoA units can each feed into the citric acid cycle.

Energy Yield From Oxidation of Stearic Acid

In Figure 10.10, we started with stearoyl CoA, an 18-carbon fatty acid attached to CoA. To determine the

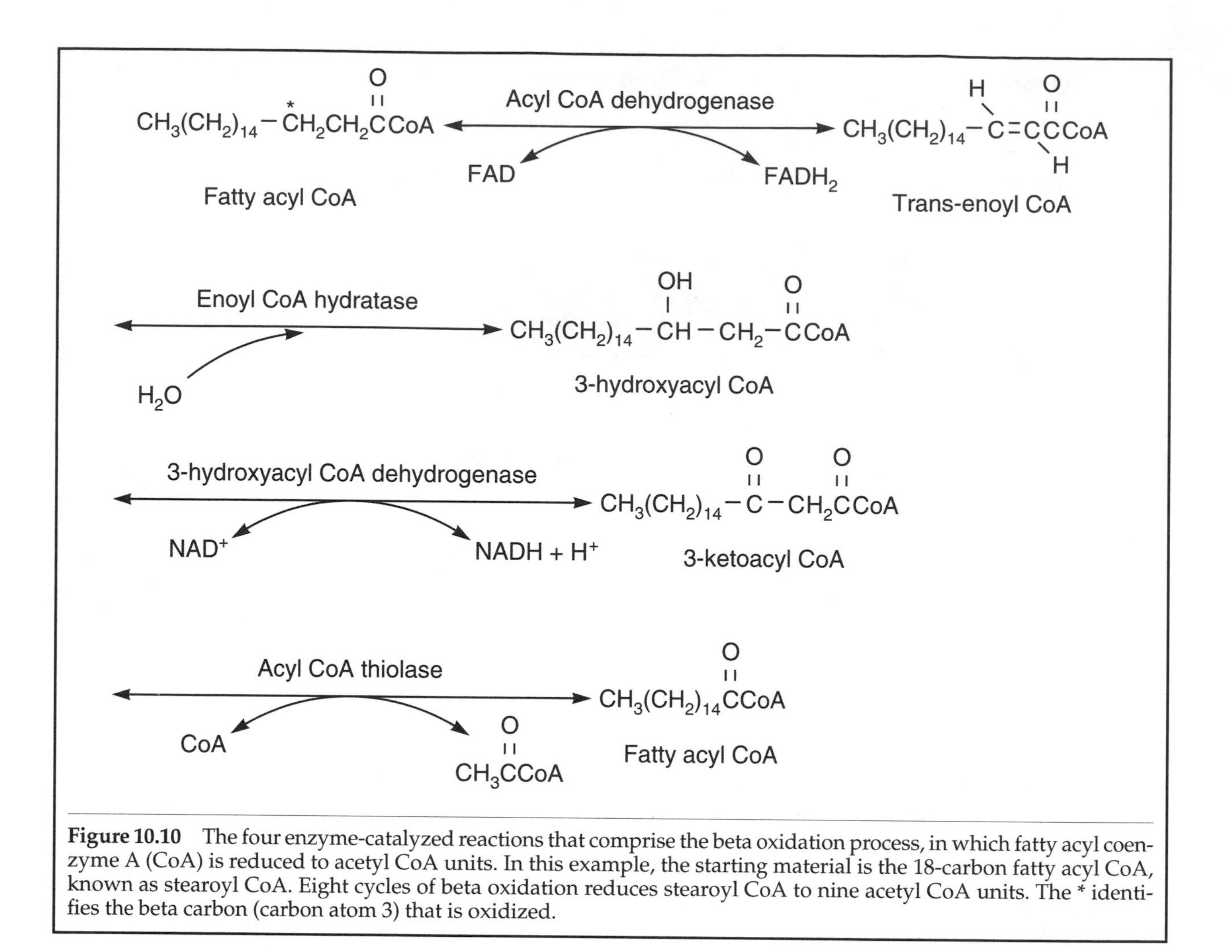

Figure 10.10 The four enzyme-catalyzed reactions that comprise the beta oxidation process, in which fatty acyl coenzyme A (CoA) is reduced to acetyl CoA units. In this example, the starting material is the 18-carbon fatty acyl CoA, known as stearoyl CoA. Eight cycles of beta oxidation reduces stearoyl CoA to nine acetyl CoA units. The * identifies the beta carbon (carbon atom 3) that is oxidized.

ATP yield beginning with just stearic acid, we get the following:

1. Formation of stearoyl CoA from stearic acid	−2 ATP
2. Formation of nine acetyl CoA from one stearoyl CoA, using the sequence of β oxidation reactions shown, yields eight FADH2 and eight NADH 1 H^+. Transfer of electrons from the eight FADH2 and eight NADH to oxygen in the electron transfer chain yields	+40 ATP
3. Oxidation of nine acetyl CoA in the CAC yields	+108 ATP
GRAND TOTAL	+146 ATP

Oxidation of Unsaturated Fatty Acids

Many fatty acids stored in body fat are unsaturated. These fatty acids are also sources of energy in the form of reduced coenzymes during β oxidation and produce acetyl CoA units. For oxidation of oleic acid, an additional step is required when the β oxidation process reaches the double bond because this carbon-to-carbon double bond is the cis configuration, whereas enoyl-CoA hydratase can only act on trans double bonds. A new enzyme, enoyl-CoA isomerase, converts the cis to a trans carbon-to-carbon double bond and β oxidation continues.

Other unsaturated fatty acids, such as linoleic acid, have their double bonds in the wrong position as well as the wrong (cis) configuration for β oxidation. For these, the enoyl-CoA isomerase converts the cis to a trans double bond, whereas an enzyme known as a reductase converts the carbon-to-carbon double bond in the wrong position into a carbon-to-carbon single bond by the addition of hydrogen.

Oxidation of Ketone Bodies

Figure 10.11 shows the three ketone bodies. D-3-hydroxybutyrate and acetoacetate are formed primarily in the matrix of liver mitochondria and to a minor extent in the kidneys. Acetoacetate undergoes slow, spontaneous decarboxylation to yield acetone and carbon dioxide. Both 3-hydroxybutyrate and acetoacetate are water soluble. A small amount of acetone is formed from acetoacetate, and because it is volatile, its presence, like that of alcohol, can be smelled in the breath.

Formation of Ketone Bodies

Ketone body formation accelerates in normal individuals when the body carbohydrate content is extremely low, as during starvation or self-controlled fasting, when extremely low carbohydrate diets are eaten, and during prolonged exercise with insufficient carbohydrate ingestion. In all of these conditions, carbohydrate content in the body is low, and thus carbohydrate utilization and blood insulin concentration are low.

Accelerated ketone body formation also occurs during uncontrolled diabetes mellitus. Although blood glucose concentration is high because insulin is lacking, carbohydrate utilization by insulin-dependent tissues is low.

With starvation or fasting, low carbohydrate diets, prolonged exercise, or uncontrolled diabetes mellitus, the adipose tissue releases large quantities of fatty acids due to an imbalance between triglyceride formation and lipolysis caused by low blood insulin concentration. Recall that insulin promotes triglyceride formation in fat cells and inhibits lipolysis. Thus the net effect of low blood insulin is that lipolysis greatly exceeds triglyceride formation, resulting in a large increase in blood free fatty acid (FFA) concentration. In uncontrolled diabetes mellitus, for example, blood FFA levels are extremely high.

Under normal conditions, the liver is able to extract about 30% of the FFA that passes through it. With high blood FFA concentration, the liver extracts even more. The fate of the extracted fatty acids by liver is formation of fatty acyl CoA, then subsequently formation of triglyceride or phospholipid, or entry into the mitochondrial matrix. During conditions favoring ketone body formation, entry of fatty acyl CoA into liver mitochondria is accelerated. β oxidation of the fatty acyl CoA is greatly augmented, forming acetyl CoA at a rate that far exceeds the capacity of the liver mitochondria to oxidize it by the citric acid cycle. Moreover, conditions favoring ketone body formation are characterized by low matrix concentrations of oxaloacetate in the liver. Accordingly, much of the acetyl CoA is directed to the formation of acetoacetate. Some acetoacetate is reduced to D-3-hydroxybutyrate. Figure 10.12 summarizes ketone body formation. The two major ketone bodies formed are acetoacetate and D-3-hydroxybutyrate. The ratio of 3-hydroxybutyrate/acetoacetate depends on the [NADH]/[NAD$^+$] ratio in the liver. Of course, acetone, the third member of the ketone body family, is formed primarily by spontaneous, nonenzymatic decarboxylation of acetoacetate.

The Fate of Ketone Bodies

In the liver, acetoacetate is a precursor for cholesterol synthesis. However, most ketone bodies are used in extrahepatic (i.e., nonliver) tissues. Ketone bodies, mainly acetoacetate and 3-hydroxybutyrate, are fuels for oxidation by mitochondria in a variety of tissues, chiefly skeletal muscle, heart, and the brain. Normally, the brain uses glucose as its primary fuel because FFA cannot pass the blood-brain barrier. However, as blood glucose concentration falls, and

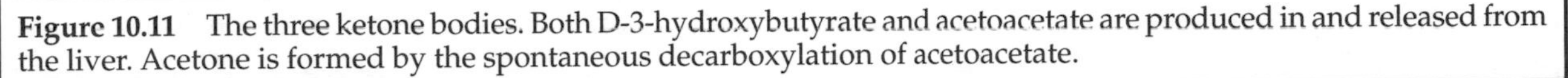

Figure 10.11 The three ketone bodies. Both D-3-hydroxybutyrate and acetoacetate are produced in and released from the liver. Acetone is formed by the spontaneous decarboxylation of acetoacetate.

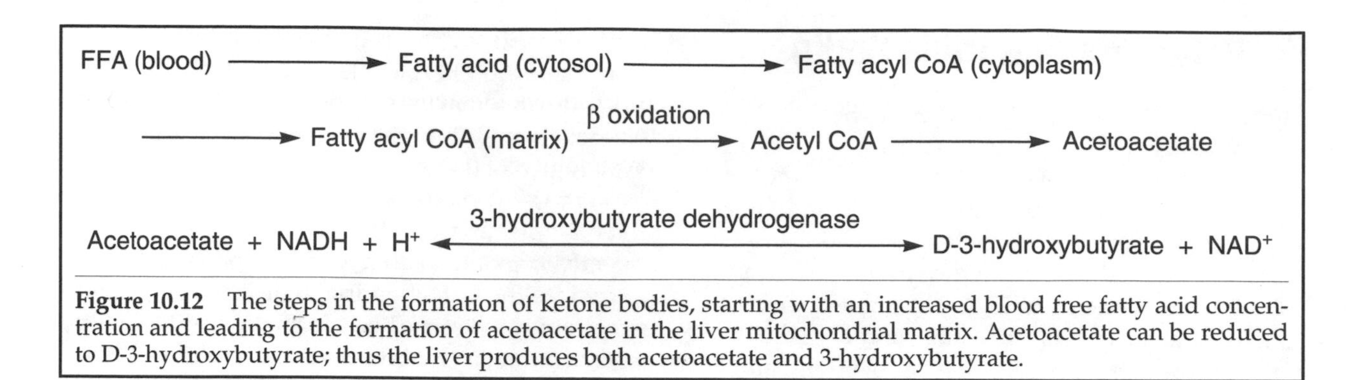

Figure 10.12 The steps in the formation of ketone bodies, starting with an increased blood free fatty acid concentration and leading to the formation of acetoacetate in the liver mitochondrial matrix. Acetoacetate can be reduced to D-3-hydroxybutyrate; thus the liver produces both acetoacetate and 3-hydroxybutyrate.

blood ketone body concentration rises, the brain can extract and use 3-hydroxybutyrate and acetoacetate. Ketone body oxidation in tissues (see Figure 10.13) occurs as follows:

1. Ketone bodies are taken up by extrahepatic cells and transported into the mitochondrial matrix.
2. D-3-hydroxybutyrate is oxidized to acetoacetate using the enzyme 3-hydroxybutyrate dehydrogenase (the reverse reaction is shown in Figure 10.12). This reaction generates NADH.
3. Acetoacetate takes a CoA from succinyl CoA and forms acetoacetyl CoA.
4. Acetoacetyl CoA changes into two acetyl CoA using the last enzyme of β oxidation of fatty acids.
5. The acetyl CoA is oxidized in the citric acid cycle.

Ketosis

Ketosis occurs during accelerated ketone body formation and is characterized by the following:

1. Ketonemia—an increase in ketone body concentration in the blood.
2. Ketonuria—loss of ketone bodies in the urine.
3. Acetone breath—loss of volatile acetone in expired air.
4. Elevated blood FFA concentration, resulting in accelerated formation of ketone bodies and low blood insulin.
5. Acidosis (or ketoacidosis)—a fall in blood pH due to the formation of H^+ when acetoacetate is formed and the loss of cations (e.g., Na^+) when ketone bodies are excreted. This can be fatal in uncontrolled diabetes mellitus.
6. Hypoglycemia (in normal persons) or hyperglycemia (in uncontrolled diabetes mellitus).

Synthesis of Fatty Acids

Most fatty acids used by humans come from dietary fat. However, humans can synthesize fatty acids from acetyl CoA, primarily in the liver. The acetyl CoA comes from amino acids, carbohydrates, and alcohol. The 16-carbon-atom palmitic acid is synthesized first and can be extended in length or desaturated to make some unsaturated fatty acids (but not the essential fatty acids needed in the diet). Synthesis of palmitic acid occurs in the cytosol, whereas oxidation occurs in the mitochondria. Synthesis and degradation of fatty acids also use different enzymes, permitting separate regulation of these two opposing processes.

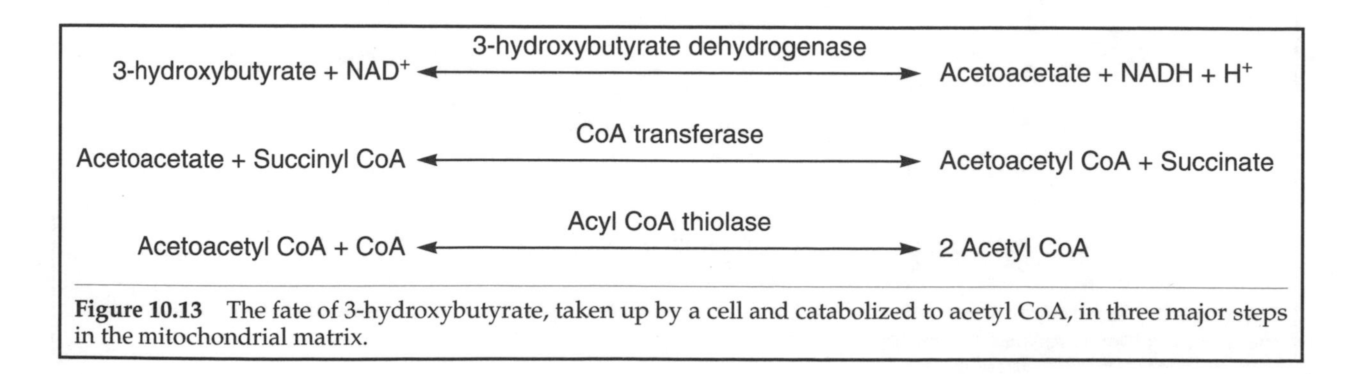

Figure 10.13 The fate of 3-hydroxybutyrate, taken up by a cell and catabolized to acetyl CoA, in three major steps in the mitochondrial matrix.

Pathway to Palmitic Acid

Synthesis starts with the carboxylation of acetyl CoA to make a three-carbon molecule known as malonyl CoA (see Figure 10.14). The enzyme acetyl CoA carboxylase catalyzes this reaction, which involves the addition of CO_2, carried to the enzyme acetyl CoA carboxylase attached to the B vitamin biotin. Although we talk of adding CO_2, the actual substrate for the acetyl CoA carboxylase is bicarbonate ion ($HCO^-{}_3$) and not CO_2. This reaction is the committed step in fatty acid synthesis and is positively affected by insulin.

Palmitic acid synthesis occurs using a large fatty acid synthase composed of two polypeptide chains. The fatty acid synthase has seven distinct enzyme activities. Although we do not have the space to cover the details of palmitic acid synthesis, Figure 10.15 shows the general process of fatty acid synthesis. To begin, an acetyl CoA enters, and the acetyl group binds to part of the fatty acid synthase. Next a malonyl CoA enters, and the malonyl group binds to another part of the fatty acid synthase.

The acetyl group then binds to the malonyl group to make a four-carbon β-keto-acyl group attached to the fatty acid synthase; a carbon dioxide molecule is eliminated. Next, three reactions remove the β-keto group: first, hydrogenation to make a hydroxy group from the keto group; second, dehydration to make a carbon-to-carbon double bond; then a further reduction to make a saturated carbon chain. Hydrogens for the two

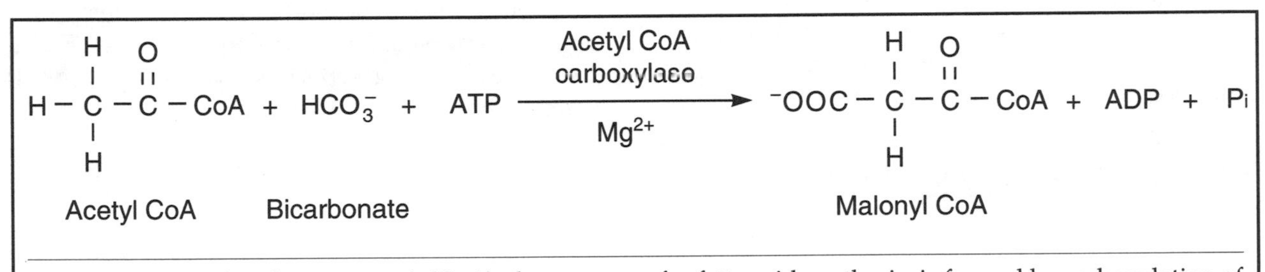

Figure 10.14 Malonyl coenzyme A (CoA), the precursor for fatty acid synthesis, is formed by carboxylation of acetyl CoA.

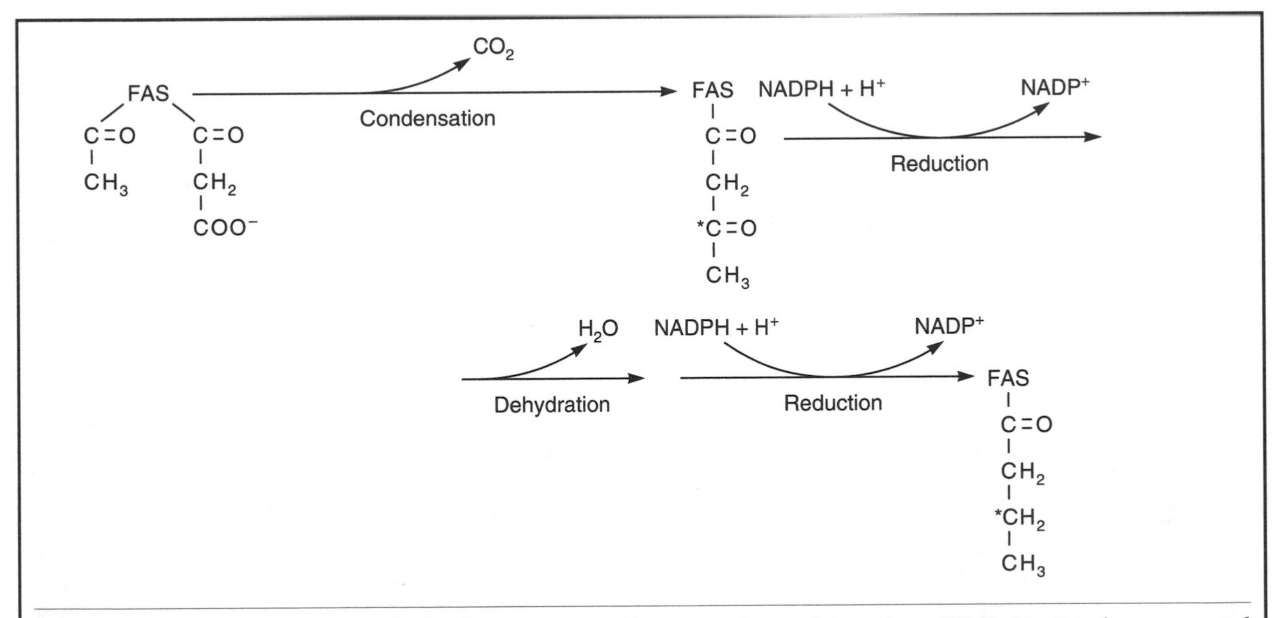

Figure 10.15 Fatty acids are synthesized on an enzyme known as fatty acid synthase (FAS). Starting from an acetyl group (from acetyl CoA) and a malonyl group (from malonyl CoA), attached at two separate sites on the FAS, the acetyl group condenses with the malonyl group and eliminates carbon dioxide. The condensation product has a β keto group (see asterisk), eliminated successively by a hydrogenation reaction using NADPH, a dehydration, and a second hydrogenation, removing the keto group and forming a four-carbon acyl group attached to the FAS. Addition of another malonyl CoA, followed by hydrogenation, dehydration, and hydrogenation generates a six-carbon fatty acyl group. This process of malonyl CoA addition and removal of the keto group continues until a 16-carbon palmityl group is cleaved from the FAS as palmitic acid.

reduction reactions come from the coenzyme NADPH, which is simply a phosphorylated derivative of NADH. Next, a new malonyl CoA molecule attaches to the four-carbon acyl group and, after elimination of a carbon dioxide, generates a new β-keto-acyl group attached to the fatty acid synthase. Another round of hydrogenation, dehydration, and hydrogenation removes the β-keto-acyl group as before.

The addition of malonyl CoA and the three-step removal of the keto group continues until the 16-carbon palmityl group is formed, still attached to the fatty acyl synthase. The palmityl group is then cleaved from the fatty acid synthase, generating palmitic acid. Note that the synthesis of fatty acids parallels β oxidation of fatty acids. Fatty acids are built and broken down two carbon atoms at a time. Three reactions remove a β-keto group during synthesis, and three introduce a β-keto group during β oxidation.

Regulation of Fatty Acid Synthesis

Fatty acids are synthesized in the liver cytosol from excess acetyl groups. These condense with oxaloacetate to form citrate under the action of citrate synthase. Citrate exits the mitochondrial matrix for the cytosol by crossing the inner mitochondrial membrane using a specific transporter. Using CoA, ATP, and an enzyme known as citrate lyase, the citrate is cleaved to oxaloacetate and acetyl CoA.

Control of fatty acid synthesis occurs through regulation of malonyl CoA formation by its enzyme acetyl CoA carboxylase. This enzyme is activated by cytosolic citrate, and its activity is inhibited by covalent phosphorylation, directed by a specific protein kinase; dephosphorylation by a protein phosphatase activates acetyl CoA carboxylase.

Malonyl CoA is also a potent inhibitor of the oxidation of fatty acids. In tissues where fatty acids can be oxidized to acetyl CoA in the mitochondria and where fatty acids can be simultaneously synthesized in the cytosol, malonyl CoA can inhibit the initial transfer of the fatty acyl group from CoA to carnitine at the cytosolic side of the inner membrane (refer to Figure 10.9). This process prevents futile cycling in which fatty acids are actively broken down in the mitochondria while at the same time being synthesized in the cytosol. Such regulation is active in the liver, the main site for fatty acid synthesis in humans.

Cholesterol

Cholesterol is an important lipid, but we only seem to hear about its bad properties. Figure 10.16 shows its chemical structure, which is not soluble in water due to its primary hydrocarbon content. In the blood, cholesterol either has a fatty acid attached to the cholesterol OH group, forming a cholesterol ester (about 70%), or is simple cholesterol (about 30%). Both cholesterol and cholesterol esters are in lipid-protein complexes called lipoproteins. Determining the concentration of cholesterol as well as cholesterol-rich lipoproteins in the blood are important clinical tests. Blood cholesterol concentration is reported three different ways: mg cholesterol/dL, mg cholesterol/L, and mM. In the U.S. the first two ways predominate, especially mg/dL; many other parts of the world use the SI system's mM units. If the molecular weight of cholesterol is 387, you should be able to convert a blood value of 150mg/dL to mM units—a value of 3.88 mM.

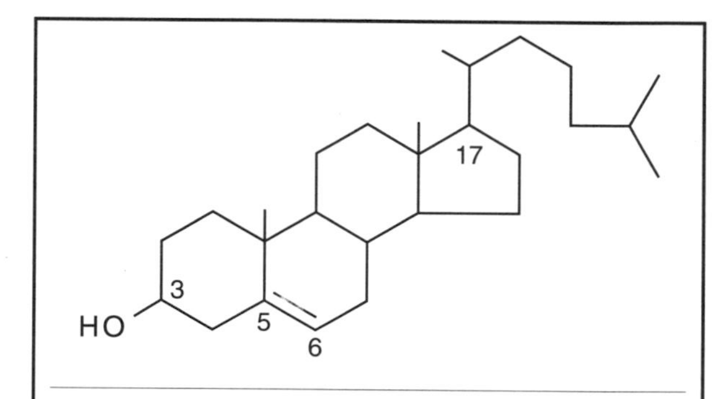

Figure 10.16 Shorthand illustration of cholesterol. The molecular formula is $C_{27}H_{47}O$. Can you find all the carbon atoms? The numbers illustrate the location of the hydroxyl (OH) group, the double bond, and the alkyl side chain.

Cholesterol is a component of membranes, a precursor for the synthesis of steroid hormones (e.g., cortisol, testosterone, estrogen, etc.), bile salts, and vitamin D, and a major component of myelin in nerves. Cholesterol is synthesized from acetyl CoA units. Although present in all cell types, cholesterol synthesis is most important in the liver, intestines, and adrenal and reproductive glands. For most people, about 60 to 70% of the body's cholesterol is synthesized; the remainder comes from the diet. Although no clear evidence exists that regular physical exercise decreases blood cholesterol levels, it may alter the type of lipoprotein carrying cholesterol in the blood.

Summary

In the body, lipids exist primarily as fatty acids, triacylglycerols (triglycerides), phospholipids, and cholesterol. Triacylglycerols consist of three long-chain fatty acids, containing saturated or unsaturated carbon-to-carbon bonds, attached by ester bonds to the three-carbon alcohol glycerol. Triacylglycerols are mainly stored in specialized cells called fat cells or adipocytes. In these cells, triacylglycerols are made by joining fatty acids to glycerol. The reverse reaction, lipolysis or lipid mobilization to yield fatty acids and glycerol, is regulated by hormone-sensitive lipase. Triacylglycerol formation is favored and lipolysis inhibited by insulin, whereas epinephrine (adrenaline), norepinephrine (noradrenaline), thyroid hormone, and growth hormone promote lipolysis.

Fatty acids released from the fat cell during lipolysis travel in the blood stream attached to the protein albumin. The fatty acids can be used by other cells as fuel. For this to happen, the fatty acid is converted to a fatty acyl CoA, then transported into the mitochondrial matrix through attachment to carnitine. In a four-step process known as beta oxidation, long-chain fatty acids are broken down into two-carbon acetyl units attached to CoA. These acetyl CoA units can then feed into the citric acid cycle. When lipolysis is increased because body carbohydrate stores are low, fatty acids can be a source of carbon by the liver to make ketone bodies. Ketone bodies are also a source of energy, but unlike fatty acids, they can be used by the brain as a fuel. Many of the problems associated with diabetes mellitus relate to uncontrolled formation of ketone bodies. Acetyl CoA units produced by the partial breakdown of excess carbohydrate or amino acids are used to make fatty acids. Unlike lipolysis, the synthesis of fatty acids occurs in the cell cytosol and involves an intermediate known as malonyl CoA.

Chapter 11

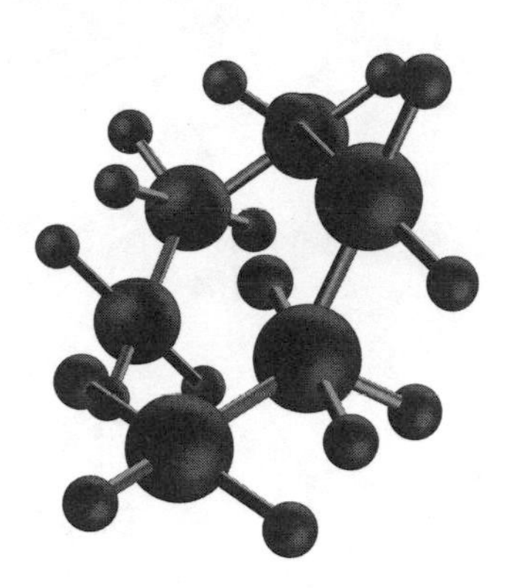

Gluconeogenesis

This section looks at gluconeogenesis—the important pathway for making glucose in the body. However, first we will examine what happens when pyruvate is oxidized in mitochondria and not converted into lactate.

Oxidation of Cytoplasmic NADH

We learned that pyruvate, a product of glycolysis, has two major fates. First, it is reduced to lactate, using the enzyme lactate dehydrogenase and NADH, which is generated in the glyceraldehyde phosphate dehydrogenase reaction and is converted back to NAD^+ when pyruvate is reduced. Second, it enters the mitochondria for terminal oxidation, using the pyruvate dehydrogenase reaction and oxidation of the resulting acetyl CoA in the citric acid cycle. If the latter occurs, the NADH formed during the glyceraldehyde phosphate dehydrogenase reaction is not oxidized by the lactate dehydrogenase reaction. If NADH is not oxidized to NAD^+, glycolysis will thus stop due to a lack of NAD^+ substrate for glyceraldehyde phosphate dehydrogenase. Also remember that cytoplasmic $[NAD^+]/[NADH]$ ratio must be high to drive glycolysis.

The simplest solution to this metabolic problem would be for cytoplasmic NADH to enter the mitochondrial matrix and be oxidized in the electron transport chain. However, the inner mitochondrial membrane is impermeable to NADH. To get around this roadblock there are two shuttle systems that transfer electrons on cytoplasmic NADH into the mitochondrion without actual inner membrane crossing by NADH.

Glycerol Phosphate Shuttle

The glycerol phosphate shuttle (see Figure 11.1) transfers electrons on cytosolic NADH to FAD, then to ubiquinone (coenzyme Q) in the mitochondrial inner membrane. The cytosolic NADH forms during the glyceraldehyde phosphate dehydrogenase reaction of glycolysis (reaction 1 in Figure 11.1). This and subsequent reactions in glycolysis are not affected by this shuttle system. In reaction 2, the cytosolic NADH, not reoxidized back to NAD^+ by the lactate dehydrogenase reaction, transfers a hydride ion and proton to dihydroxyacetone phosphate, changing it to glycerol 3-P and oxidizing NADH to NAD^+. The enzyme catalyzing this reaction (2 in Figure 11.1) is cytosolic glycerol phosphate dehydrogenase. Glycerol 3-P diffuses to the outer side of the inner mitochondrial membrane and is oxidized by mitochondrial glycerol phosphate dehydrogenase (reaction 3). This enzyme is located in the inner membrane but faces the intermembranous space so that glycerol 3-P need not

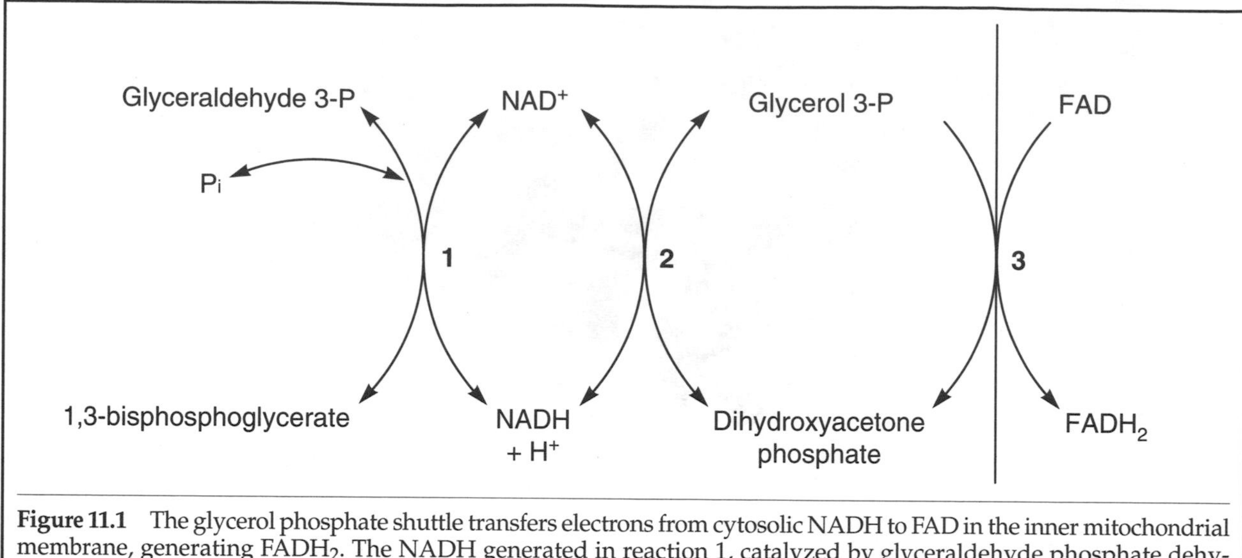

Figure 11.1 The glycerol phosphate shuttle transfers electrons from cytosolic NADH to FAD in the inner mitochondrial membrane, generating $FADH_2$. The NADH generated in reaction 1, catalyzed by glyceraldehyde phosphate dehydrogenase, is used to reduce dihydroxyacetone phosphate in reaction 2, catalyzed by cytoplasmic glycerol phosphate dehydrogenase. Glycerol 3-P diffuses to the inner mitochondrial membrane and transfers a pair of electrons to FAD in reaction 3, catalyzed by a mitochondrial glycerol phosphate dehydrogenase. Electrons on $FADH_2$ are subsequently transferred to ubiquinone. The dihydroxyacetone phosphate created in reaction 3 is now available to accept another pair of electrons from NADH. The shuttle is irreversible because reaction 3 is irreversible.

penetrate the inner membrane. Instead of using the NAD^+ coenzyme, the mitochondrial form of glycerol phosphate dehydrogenase uses FAD. FAD binds to the mitochondrial form of the glycerol phosphate dehydrogenase and is one of the other flavoprotein dehydrogenases described as complex II in oxidative phosphorylation. In reaction 3, the products are $FADH_2$ and dihydroxyacetone phosphate. The electrons on $FADH_2$ transfer to ubiquinone (coenzyme Q) in the electron transport chain and then to oxygen. The dihydroxyacetone phosphate then diffuses back to again accept electrons from cytoplasmic NADH.

The glycerol phosphate shuttle exhibits no net consumption or production of dihydroxyacetone phosphate or glycerol 3-P. It simply transfers electrons from cytosolic NADH to the electron transport chain. This shuttle will generate two ATP per pair of electrons transferred from cytosolic NADH to oxygen and is irreversible because reaction 3 goes in only one direction. Although not the most prevalent of the two major shuttle systems, it is easier to explain. The glycerol phosphate shuttle operates to a minor extent in a variety of tissues, such as brain but is very important in fast twitch (type II) skeletal muscle fibers.

Malate-Aspartate Shuttle

The malate-aspartate shuttle is more complicated and also irreversible. It is the dominant shuttle for the liver, the heart, and slow twitch (type I) muscle fibers. This shuttle system uses NADH reduction in the cytosol and NADH oxidation in the mitochondria (as opposed to $FADH_2$ oxidation in the glycerol phosphate shuttle). As a result, for each two cytosolic electrons on NADH transferred to oxygen in the mitochondria, we get three ATP. Although not shown, this shuttle involves the movement of four substances across the inner membrane.

ATP Yield From the Oxidation of Glucose

Table 11.1 summarizes, for the two dominant shuttle systems for transferring cytosolic electrons (reducing equivalents) into the mitochondrion, the number of moles of ATP generated by the complete oxidation of glucose.

Overview of Gluconeogenesis

Gluconeogenesis, as the name suggests, means new glucose formation from noncarbohydrate precursors, such as lactate, pyruvate, glycerol, and the carbon skeletons of amino acids, left when the amino group is removed. Because the body cannot oxidize amino groups on amino acids, it converts these primarily to urea, which is excreted in the urine (Chapter 13 describes this process). Gluconeogenesis takes place

Table 11.1 ATP Molecules Formed During the Complete Oxidation of Glucose in a Cell Using the Glycerol Phosphate or the Malate-Aspartate Shuttle

Reaction	Number of ATP
Hexokinase reaction	−1
Phosphofructokinase reaction	−1
Phosphoglycerate kinase reaction	2 *
Pyruvate kinase reaction	2 *
Oxidation of cytoplasmic NADH—	
glycerol phosphate shuttle	4 *
or	
malate-aspartate shuttle	6 *
Pyruvate dehydrogenase reaction	6 *
Oxidation of acetyl CoA, including electron transport chain	24 *
Grand total	36 or 38

Note. The asterisk (*) means that each glucose molecule yields two ATP because each glucose generates two 3-carbon units (e.g., pyruvate) and two acetyl CoA.

mainly in the liver (about 90%) and in the kidney cortex (10%).

A number of tissues are obligatory glucose users. The central nervous system (principally the brain) needs about 125 g of glucose each day. Other tissues need an additional 30 to 40 g of glucose a day. Thus glucose-dependent tissues need about 160 g of glucose. If we exercise or do work, we also use blood glucose in the contracting muscles, and fat cells need glucose as a source of glycerol 3-phosphate to make triglyceride. The body thus has a constant need for glucose.

Gluconeogenesis becomes important whenever dietary carbohydrate is inadequate to supply the body sufficient glucose, that is, during fasting or starvation, low carbohydrate diets, or prolonged exercise. Gluconeogenesis also occurs whenever blood lactate concentration increases, as during moderate to severe exercise. Although lactate is used as a fuel by the heart or non-exercising muscles or to make glycogen in muscle, a considerable amount is extracted by the liver to make glucose. In fact, in the Cori Cycle, carbon transfers from liver to muscle as glucose and from muscle to liver as lactate.

As one might expect, conditions in which gluconeogenesis becomes important exhibit an altered hormone profile. Table 11.2 outlines three hormones important for gluconeogenesis.

Reactions of Gluconeogenesis

New glucose is primarily made from simple precursors by the reversal of the glycolytic pathway, as Figure 11.2 reveals. Most glycolytic reactions can go backward; however, three are irreversible: the pyruvate kinase, the phosphofructokinase, and the hexokinase/glucokinase reactions. Glycolysis is thus not completely reversible because these reactions act as one-way valves. However, means for bypassing the irreversible reactions exist.

Focus on pyruvate in Figure 11.2. If glycolysis is to be reversed, we immediately have a problem getting from pyruvate to phosphoenolpyruvate because the pyruvate kinase reaction is irreversible. To bypass this reaction, pyruvate is converted to oxaloacetate and oxaloacetate to phosphoenolpyruvate (PEP). These conversions require two reactions with new enzymes, pyruvate carboxylase and phosphoenolpyruvate carboxykinase

$$\text{ATP} + \text{pyruvate} + CO_2 + H_2O \xrightarrow[Mg^{2+}]{\text{pyruvate carboxylase}} \text{oxaloacetate} + \text{ADP} + \text{Pi}$$

The pyruvate carboxylase reaction takes place in the mitochondrial matrix of the liver in two steps.

Table 11.2 The Major Hormones in Gluconeogenesis

Hormone	Source	Details
Glucagon	α-cells of the pancreas	Secreted when blood glucose falls below normal. Stimulates breakdown of liver glycogen and gluconeogenesis.
Insulin	β-cells of the pancreas	Secreted in response to increase in blood glucose. Suppresses gluconeogenesis.
Cortisol	Adrenal cortex	Released during prolonged exercise or other conditions of low body carbohydrate stores. Decreases protein synthesis and increases protein catabolism, thereby increasing amino acids for gluconeogenesis.

The first involves the carboxylation of biotin, the tightly bound coenzyme of pyruvate carboxylase. The actual substrate is bicarbonate, the source of the carbon dioxide. Step two involves the transfer of carbon dioxide on biotin to pyruvate, generating oxaloacetate.

Pyruvate carboxylase is a regulated enzyme requiring acetyl CoA to activate it, although the acetyl CoA does not participate in the reaction. It simply binds to an allosteric site on pyruvate carboxylase and makes it active. An increase in acetyl CoA can signal an overactive β oxidation or that there are too few citric acid cycle intermediates in the mitochondrial matrix. Because the pyruvate carboxylase reaction produces the citric acid cycle intermediate oxaloacetate, it produces all the intermediates preceding oxaloacetate, if these reactions are reversible.

The other reaction needed to bypass the irreversible pyruvate kinase reaction is one that converts oxaloacetate to phosphoenolpyruvate.

$$\text{oxaloacetate} + \text{GTP} \xleftrightarrow[\text{Mg}^{2+}]{\text{PEP carboxykinase}} \text{phosphoenolpyruvate} + \text{GDP} + CO_2$$

The phosphoenolpyruvate (PEP) carboxykinase reaction is reversible and can occur in both the mitochondrial matrix and the cytosol in humans. The first step is decarboxylation, followed by phosphorylation using GTP, which produces the energy-rich PEP. No biotin is involved. Note that it costs the equivalent of two ATP to bypass the irreversible pyruvate kinase reaction.

Once the phosphoenolpyruvate is formed, the glycolytic sequence can reverse to fructose, 1,6-bisphosphate (in Figure 11.2, fructose 1,6-P_2). To get to glucose, the fructose, 1,6-bisphosphate must be converted to fructose 6-P. Because the phosphofructokinase reaction is irreversible, fructose 1,6-bisphosphatase hydrolyzes the phosphate group on the 1-carbon of fructose, producing fructose 6-P in the following equation:

$$\text{fructose 1,6-}P_2 + H_2O \xrightarrow{\text{fructose 1,6-bisphosphatase}} \text{fructose 6-P} + \text{Pi}$$

Phosphatases, as you have seen, work the opposite way to kinases in that they remove phosphate groups from their substrate. The fructose 6-P is converted to glucose 6-P by the glycolytic enzyme glucose phosphate isomerase. At glucose 6-P we are again confronted with a problem because hexokinase and glucokinase catalyze irreversible reactions. Free glucose is thus obtained from glucose 6-P through the use of another phosphatase, glucose 6-phosphatase. The new glucose molecules formed by gluconeogenesis can now be exported to the blood to maintain its sugar level.

$$\text{glucose 6-P} + H_2O \xrightarrow{\text{glucose 6-phosphatase}} \text{glucose} + \text{Pi}$$

Figure 11.2 also shows the role of glycerol, which, after released from fat cells during lipolysis, can be taken up by the liver from the blood. It is first phosphorylated by glycerol kinase to make glycerol 3-P, which is oxidized to form dihydroxyacetone phosphate. The enzyme that catalyzes this reaction is the cytosolic form of glycerol 3-P dehydrogenase that we saw in the glycerol phosphate shuttle. During the oxidation of glycerol 3-P, NAD^+ is reduced to NADH. During starvation, glycerol produced during lipolysis of stored fat can act as a significant gluconeogenic precursor, providing a third or more of the glucose produced by the liver.

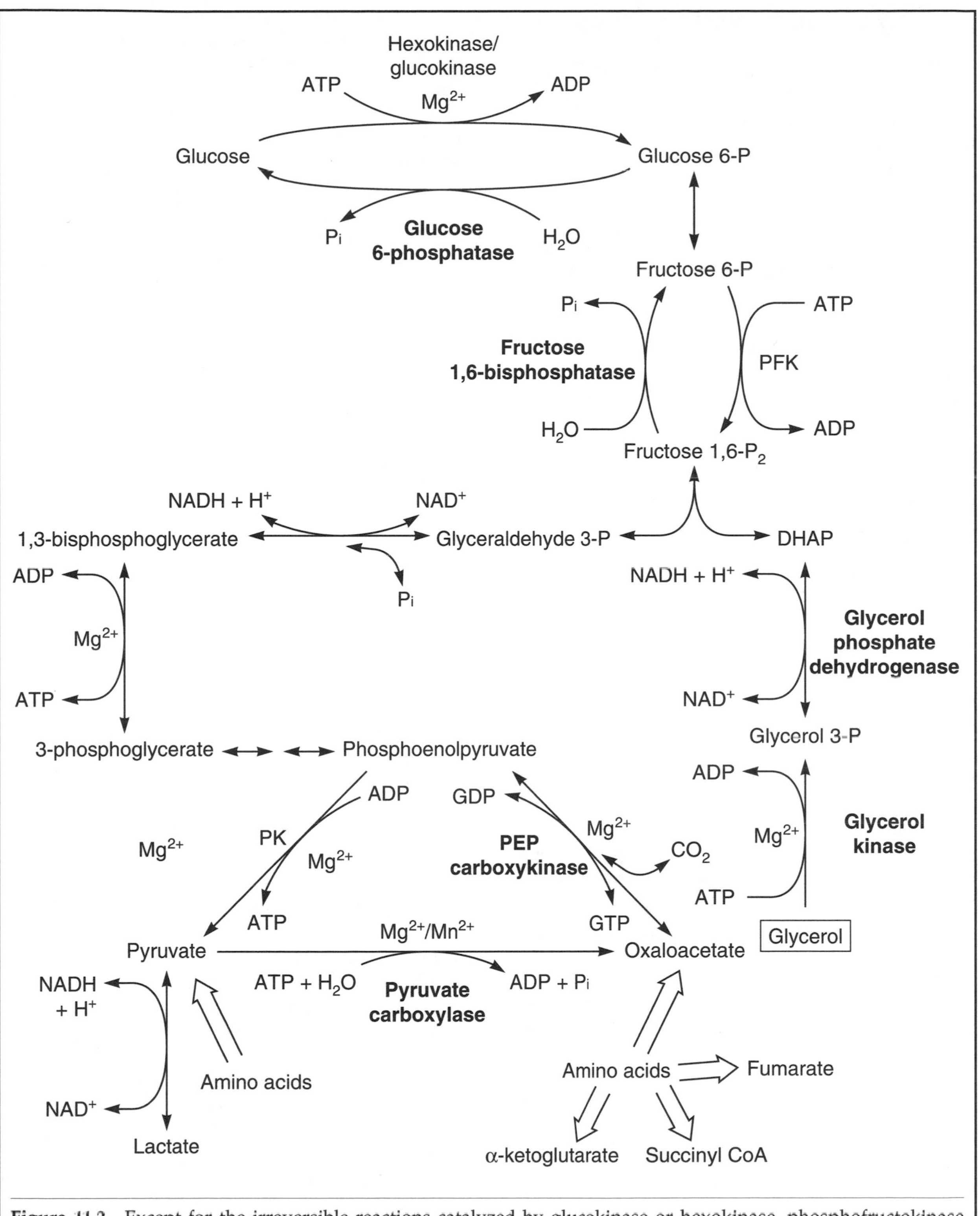

Figure 11.2 Except for the irreversible reactions catalyzed by glucokinase or hexokinase, phosphofructokinase (PFK), and pyruvate kinase (PK), glycolysis in the liver could go backward from lactate to glucose. New reactions, catalyzed by the bolded enzymes, allow glycerol, released from fat cells, and amino acid carbon skeletons to be used as a source of new glucose. DHAP is dihydroxyacetone phosphate, and PEP phosphoenolpyruvate.

The Role of Amino Acids

The last chapter presents a comprehensive discussion of amino acid metabolism. For now, a simple overview (refer again to Figure 11.2) demonstrates that amino acids can be used as a source of carbon atoms to make glucose. In fact, all or part of the carbon atoms of 18 of the 20 common amino acids help form glucose; we call these glucogenic amino acids. In mammalian tissues the exceptions are leucine and lysine, which are ketogenic amino acids. Because considerable glucose comes from their carbon skeletons, amino acids are not wasted but become fuel.

Before the carbon atoms of the glucogenic amino acids can form glucose, the amino groups must be removed. One way of removing them is to transfer them to other molecules. Figure 11.3 shows such a transfer for the branched-chain amino acids (leucine, isoleucine, and valine). The amino groups are transferred primarily in muscle, to a α-ketoglutarate to make branched-chain keto acids and glutamic acid. The amino group on glutamic acid is then transferred to pyruvate, regenerating the α-ketoglutarate and forming alanine. In Figure 11.3, enzymes known as transaminases (also called aminotransferases) transfer the amino group from an amino acid to a keto acid, making a new keto and amino acid. Reactions catalyzed by transaminases, which use vitamin B_6 as a cofactor, are known as transamination reactions.

Alanine formed in skeletal muscle is released to the blood. Liver then transfers the amino group from alanine to α-ketoglutarate, making glutamic acid, and the carbon skeleton remaining, that is, pyruvate, is used to make glucose by gluconeogenesis. Figure 11.4 shows the cycling of amino groups from muscle to liver by way of alanine and the formation of glucose. This has been described as the glucose-alanine cycle, and during exercise the rate of this cycle is increased.

Amino groups from amino acids are used to make urea in the urea cycle, which occurs in the liver. The urea is then excreted in the urine. When most amino groups are removed, citric acid cycle intermediates such as oxaloacetate, α-ketoglutarate, fumarate, succinyl CoA, and pyruvate result. These intermediates and pyruvate can be used to make glucose, as outlined in Figure 11.2. The glucose-alanine cycle thus illustrates the importance of amino acids as a source of glucose. It also shows the relative importance of certain tissues in the body. Muscle, for example, the primary protein-containing tissue, is sacrificed to make glucose when food energy is insufficient.

Regulation of Gluconeogenesis

Gluconeogenesis occurs primarily in the liver, with most of the reactions in the cytosol. Liver cytosol also contains enzymes that catalyze glycolysis. Because much of gluconeogenesis occurs by reversal of glycolytic reactions, simultaneous activity of both glycolysis and gluconeogenesis would be inefficient. To prevent this, glycolysis and gluconeogenesis are reciprocally regulated, aided by two hormones. Insulin stimulates glycolysis, and glucagon stimulates gluconeogenesis. During conditions favoring liver glycolysis (after eating, for example), blood insulin is increased, whereas glucagon is decreased. During conditions favoring gluconeogenesis, insulin concentration is low, whereas glucagon is high. Glucagon stimulates gluconeogenesis by increasing liver cell cAMP concentration. Elevated cAMP activates an enzyme that phosphorylates pyruvate kinase (the glycolytic enzyme) and thus inactivates pyruvate kinase. It also decreases the concentration of fructose 2,6-bisphosphate (F 2,6-P_2), a stimulator of phosphofructokinase and an inhibitor of fructose bisphosphatase.

Figure 11.5 shows the major sites for this reciprocal regulation, the phosphofructokinase and fructose bisphosphatase reactions. Control is essential because, if both enzymes are active simultaneously, a futile

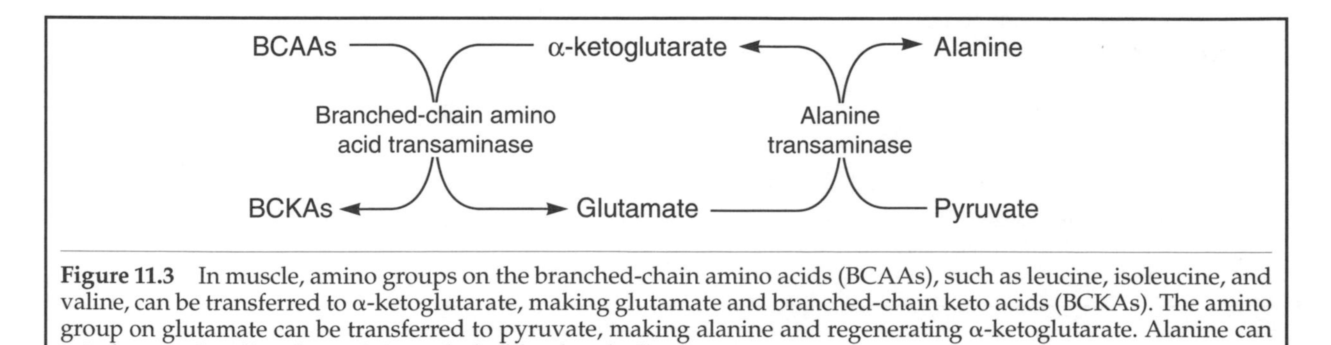

Figure 11.3 In muscle, amino groups on the branched-chain amino acids (BCAAs), such as leucine, isoleucine, and valine, can be transferred to α-ketoglutarate, making glutamate and branched-chain keto acids (BCKAs). The amino group on glutamate can be transferred to pyruvate, making alanine and regenerating α-ketoglutarate. Alanine can exit the muscle cell and travel through the blood to the liver.

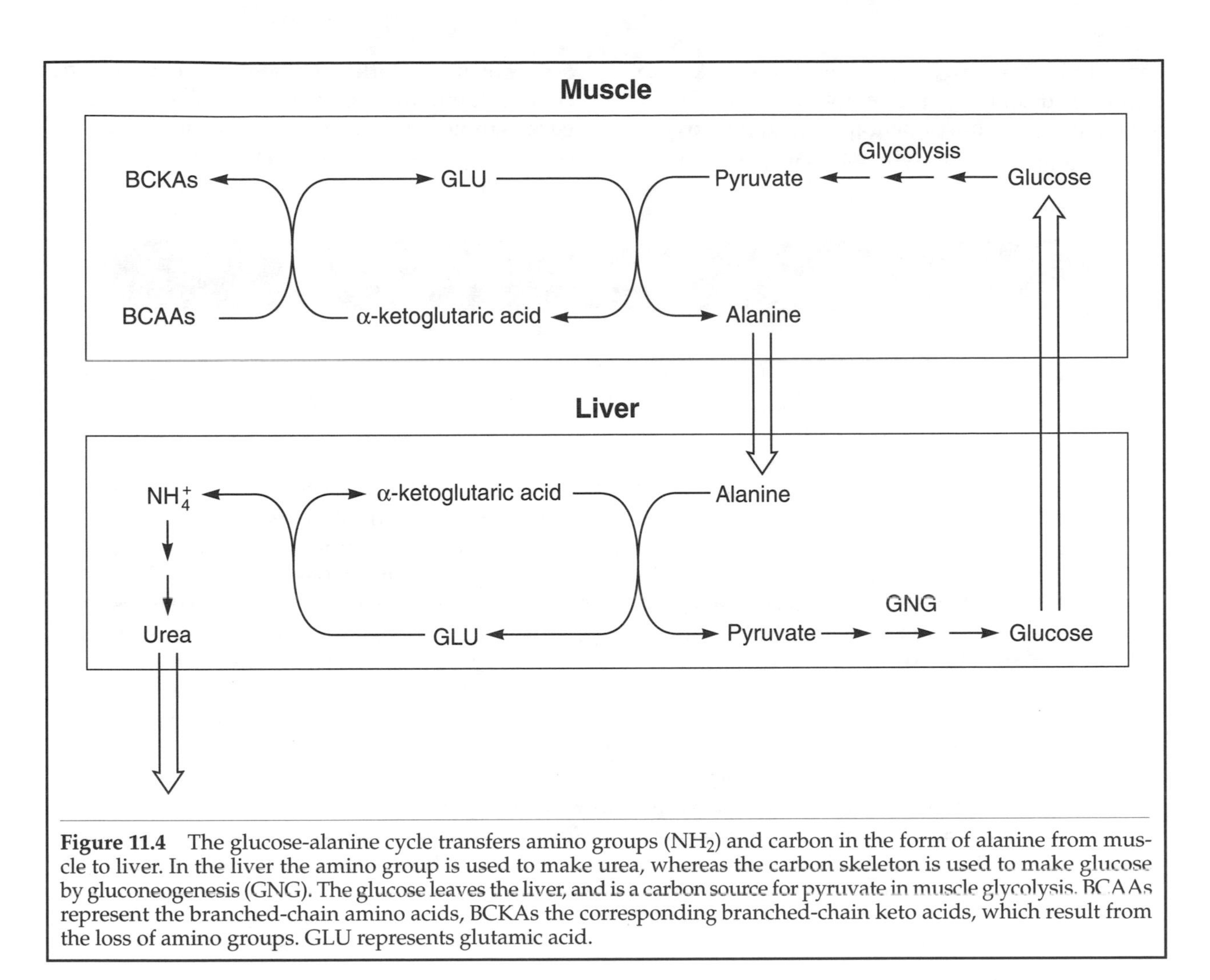

Figure 11.4 The glucose-alanine cycle transfers amino groups (NH_2) and carbon in the form of alanine from muscle to liver. In the liver the amino group is used to make urea, whereas the carbon skeleton is used to make glucose by gluconeogenesis (GNG). The glucose leaves the liver, and is a carbon source for pyruvate in muscle glycolysis. BCAAs represent the branched-chain amino acids, BCKAs the corresponding branched-chain keto acids, which result from the loss of amino groups. GLU represents glutamic acid.

cycle of phosphorylation of fructose 6-P and dephosphorylation of fructose 1,6-bisphosphate results, with nothing accomplished except the hydrolysis of ATP. Therefore, when phosphofructokinase is active, fructose 1,6-bisphosphatase is inactive. Allosteric effectors that activate phosphofructokinase inhibit fructose 1,6-biphosphatase and vice versa, thus responding to the needs of the body. In addition, acetyl CoA activates pyruvate carboxylase, and the matrix content of acetyl CoA will be elevated when conditions favor gluconeogenesis. Also, under gluconeogenic conditions, for example, especially stressful situations, blood cortisol increases. We saw previously that cortisol provides amino acids as gluconeogenic precursors.

Gluconeogenic control can be rapid and short-term. However, regulating the expression of genes for the regulatory enzymes of glycolysis and gluconeogenesis can play a longer term role in dictating the activities of glycolysis and gluconeogenesis in the liver. For example, starvation or a low carbohydrate diet (with low insulin concentration) are

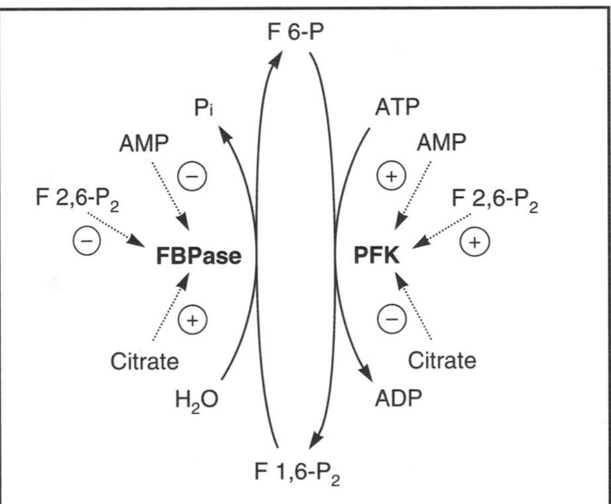

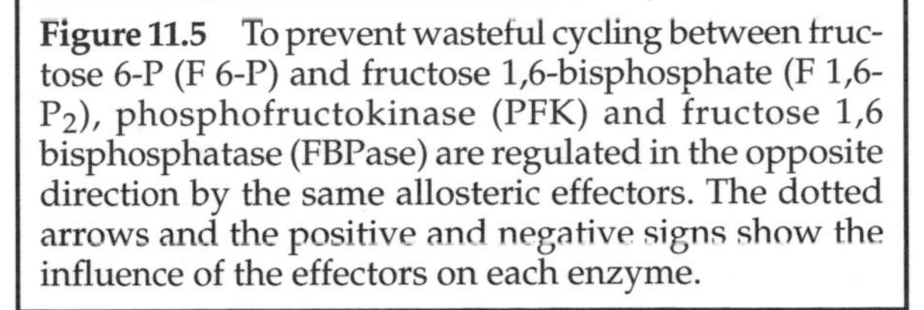
Figure 11.5 To prevent wasteful cycling between fructose 6-P (F 6-P) and fructose 1,6-bisphosphate (F 1,6-P_2), phosphofructokinase (PFK) and fructose 1,6 bisphosphatase (FBPase) are regulated in the opposite direction by the same allosteric effectors. The dotted arrows and the positive and negative signs show the influence of the effectors on each enzyme.

conditions that repress the genes for glucokinase, phosphofructokinase, and pyruvate kinase and activate those for phosphoenolpyruvate carboxykinase, fructose 1,6 bisphosphatase, and glucose 6-phosphatase. With adequate food energy, especially carbohydrate, glycolytic regulatory enzyme gene activity is increased and gluconeogenic regulatory enzyme gene activity is repressed.

Summary

When pyruvate is oxidized in the mitochondrion and not reduced to lactate during glycolysis, NADH can potentially build up to the extent that there is too little NAD^+ for the glyceraldehyde phosphate dehydrogenase reaction. Oxidation of cytoplasmic NADH, other than by lactate dehydrogenase, occurs through the action of two shuttle systems. The glycerol phosphate shuttle and the malate-aspartate shuttle transfer electrons from the cytosol to the mitochondria for use by the electron transfer chain. When body carbohydrate content is decreased, glucose, an essential fuel, is made from a variety of noncarbohydrate sources, such as lactate, pyruvate, and the carbon skeletons of amino acids. This process, gluconeogenesis, takes place in liver and kidney and involves the reversible reactions of glycolysis. New reactions that bypass irreversible glycolytic reactions are catalyzed by pyruvate carboxylase, phosphoenolpyruvate carboxykinase, fructose 1,6-bisphosphatase, and glucose 6-phosphatase. These reactions allow the liver to produce glucose during times of need while glycolysis is virtually shut down. The pancreatic hormone glucagon promotes gluconeogenesis. Glycerol, produced when triacylglycerols are broken down, is an important gluconeogenic precursor in liver. Also when the amino group of the 20 common amino acids is removed, the carbon skeletons of 18 of these can be converted to glucose. The fact that skeletal muscle can be severely wasted during diets providing inadequate food energy demonstrates the importance of the brain as a tissue since it can be a significant user of glucose derived from amino acids obtained from muscle proteins.

Chapter 12

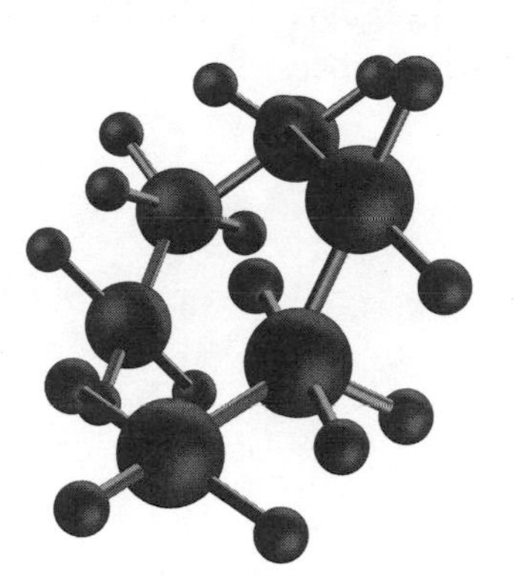

Glycogen Metabolism

Glycogen is a polysaccharide, composed of hundreds of glucose molecules (monosaccharides) joined end to end, with prevalent branches.

Glycogen Storage

The major stores of glycogen are in liver and skeletal muscle. Table 12.1 provides approximate values for the amount of glycogen in liver and skeletal muscle for men and women under three different dietary conditions. As shown, the amount of carbohydrate in the diet influences the amount of stored glycogen. The normal, mixed diet commonly eaten by North American people contains about 45 to 50% carbohydrate. Glycogen is stored in both liver and muscle following a meal. However, after exercise, when muscle glycogen levels are reduced, glycogen is stored preferentially in the exercised muscles.

Mechanism of Glycogen Storage

In glycogenesis, glucose units are added one at a time to existing glycogen molecules, creating long unbranched chains. A branching enzyme then creates the highly-branched final structure. Glucose enters the cell and is phosphorylated to glucose 6-P by hexokinase (muscle) or glucokinase (liver). Figure 12.1 illustrates the three-step process of glycogen formation, starting from glucose 6-phosphate. Overall, glycogen synthesis is irreversible, due in part to subsequent hydrolysis of inorganic pyrophosphate (PPi) by inorganic pyrophosphatase. The glycogen synthase reaction is also irreversible. Glycogen synthase produces long chains of glucose molecules that the branching enzyme transforms into the tree-like structure of glycogen found *in vivo*. The UTP needed to make glycogen comes from the nucleoside diphosphate kinase reaction.

$$\text{UDP} + \text{ATP} \xleftrightarrow[\text{Mg}^{2+}]{\text{nucleoside diphosphate kinase}} \text{UTP} + \text{ADP}$$

Mechanism of Glycogen Breakdown

Glycogenolysis, or the breakdown of glycogen, is the phosphorolytic cleavage of glucose units, one at a time, from glycogen molecules through the introduction of Pi. Figure 12.2 shows the process of glycogenolysis. Glycogen phosphorylase, often simply called phosphorylase, acts on chains of glucose units and a debranching enzyme is necessary to remove branches.

In liver, glycogenolysis mainly provides glucose for the blood. Therefore, glucose 6-phosphate from the phosphoglucomutase reaction is most likely dephosphorylated to free glucose, although it may

Table 12.1 Approximate Glycogen Stores in Liver and Muscle for Adults Following Normal, High Carbohydrate (CHO), and Low CHO Diets

Storage site	Tissue weight (kg)	Total glycogen content (g)		
		Normal diet	High CHO diet	Low CHO diet
Man, 70 kg				
Liver	1.2	40-50	70-90	0-20
Muscle	32.0	350	600	250
Woman, 55 kg				
Liver	1.0	35-45	60-70	0-15
Muscle	22.0	242	410	170

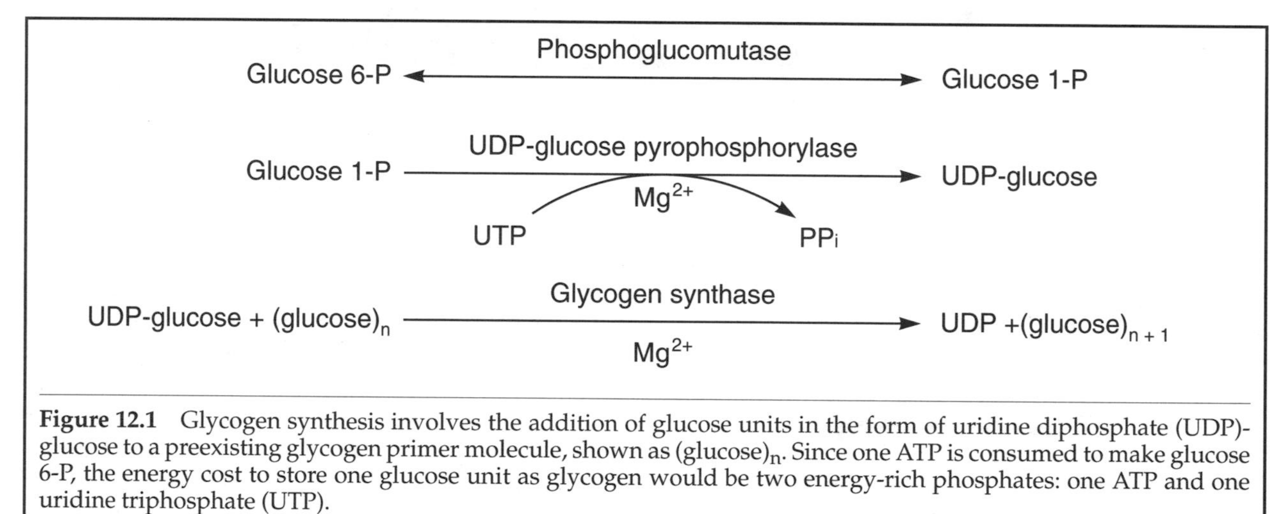

Figure 12.1 Glycogen synthesis involves the addition of glucose units in the form of uridine diphosphate (UDP)-glucose to a preexisting glycogen primer molecule, shown as $(glucose)_n$. Since one ATP is consumed to make glucose 6-P, the energy cost to store one glucose unit as glycogen would be two energy-rich phosphates: one ATP and one uridine triphosphate (UTP).

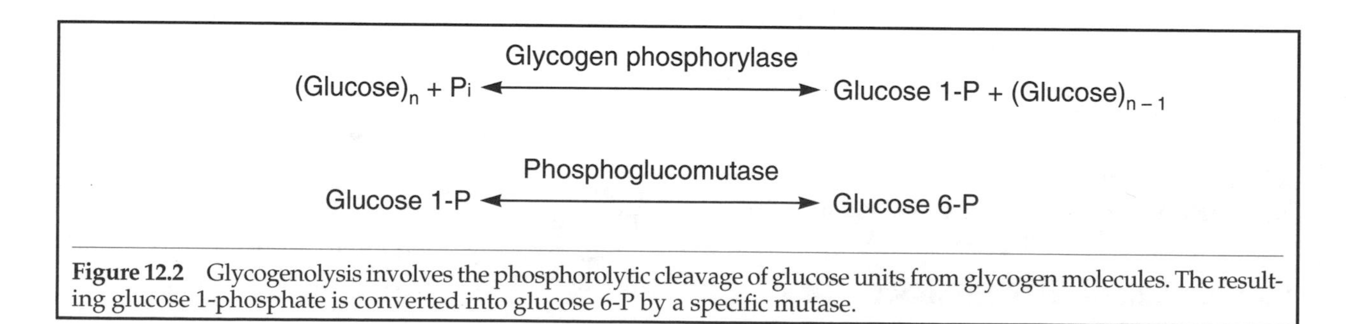

Figure 12.2 Glycogenolysis involves the phosphorolytic cleavage of glucose units from glycogen molecules. The resulting glucose 1-phosphate is converted into glucose 6-P by a specific mutase.

also be used in the glycolytic sequence to make pyruvate. In muscle, glycogen breakdown generates glucose 6-phosphate units for glycolysis.

Regulation of Glycogen Metabolism

In liver and muscle cytoplasm are glycogen particles that contain glycogen as well as enzymes both to make and break down glycogen. If the processes of synthesis and breakdown are simultaneously active, a futile cycle results that does nothing except use energy-rich phosphates such as UTP. Thus glycogen phosphorylase should be active when glycogen synthase is inactive or vice versa.

In the liver, phosphorylase should be inactive following a meal, but glycogen synthase should be active to store the glucose obtained from food. Between meals, liver phosphorylase should be active to provide glucose for the blood, whereas glycogen synthase should be inactive. In rested muscle, synthase should be active and phosphorylase inactive following a meal, but if the muscle starts to work, the phosphorylase should be active and the synthase inactive.

In fact, we might expect the activity of phosphorylase to be graded during exercise; its activity should be most active during very hard exercise when carbohydrate is the most needed fuel and least active when exercise intensity is low enough for oxidation of fatty acids to maintain ATP levels in the exercising muscle. Control of glycogen phosphorylase can thus be tied to the whole strategy of using carbohydrate for oxidative phosphorylation when there is plenty available or when the muscle demand for ATP can only be satisfied by carbohdrate degradation.

Regulation by Enzyme Phosphorylation

One enzyme can be active and the other simultaneously inhibited if the two respond in opposite directions to the same stimulus. The main regulation of these enzymes is covalent attachment of a phosphate group, or phosphorylation, which requires a protein kinase. To remove the phosphate groups requires a protein phosphatase. Figure 12.3 outlines this process. The E represents the unphosphorylated form of either phosphorylase or glycogen synthase, whereas E-P represents the phosphorylated form. A kinase phosphorylates the enzymes, and a phosphatase dephosphorylates them. Phosphorylation activates phosphorylase and dephosphorylation inactivates it. In contrast, phosphorylation inactivates glycogen synthase but dephosphorylation activates it.

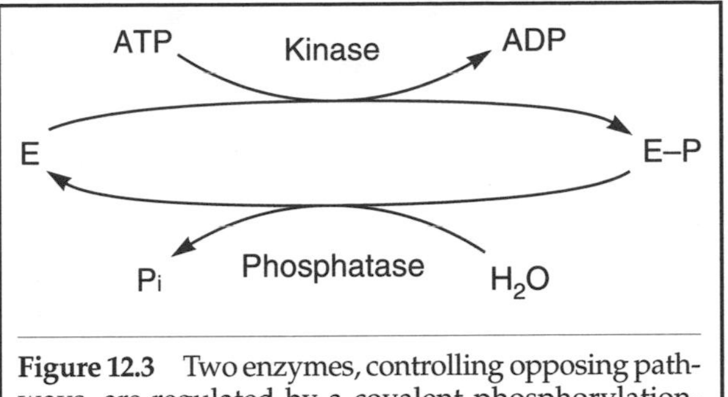

Figure 12.3 Two enzymes, controlling opposing pathways, are regulated by a covalent phosphorylation-dephosphorylation system. For glycogen metabolism, phosphorylation of glycogen phosphorylase increases activity, whereas phosphorylation of glycogen synthase decreases activity. Dephosphorylation of the two enzymes decreases phosphorylase activity and increases synthase activity.

Regulation of Phosphorylase in Muscle

The unphosphorylated form of phosphorylase is known as phosphorylase b, which is normally inactive. The phosphorylated form is phosphorylase a; it is active in breaking down glycogen. Figure 12.4 shows how this process is controlled.

Phosphorylation converts phosphorylase b (phos b) to phosphorylase a (phos a). The phosphorylation is catalyzed by phosphorylase kinase, which exists in two forms. The dephosphorylated form is inactive unless it binds four calcium ions (Ca^{2+}). The cytosolic [Ca^{2+}] increases when a nerve activates a muscle fiber, causing the release of calcium ions from the sarcoplasmic reticulum. The phosphorylated form of phosphorylase kinase catalyzes the phosphorylation of phosphorylase b. Phosphorylase kinase becomes active through phosphorylation by a cAMP-dependent protein kinase catalytic subunit, shown as A kinase. As previously discussed, cAMP forms when adenylate cyclase is activated, for example, in muscle this comes about when epinephrine (adrenaline) binds to its receptor on the cell membrane. The blood epinephrine concentration increases during exercise or even the anticipation of exercise or competition. Figure 12.4 shows that protein phosphatase-1 also catalyzes the dephosphorylation of both phosphorylase kinase and phosphorylase a.

In addition to the above phosphorylation-dephosphorylation mechanisms, both phosphorylase a and

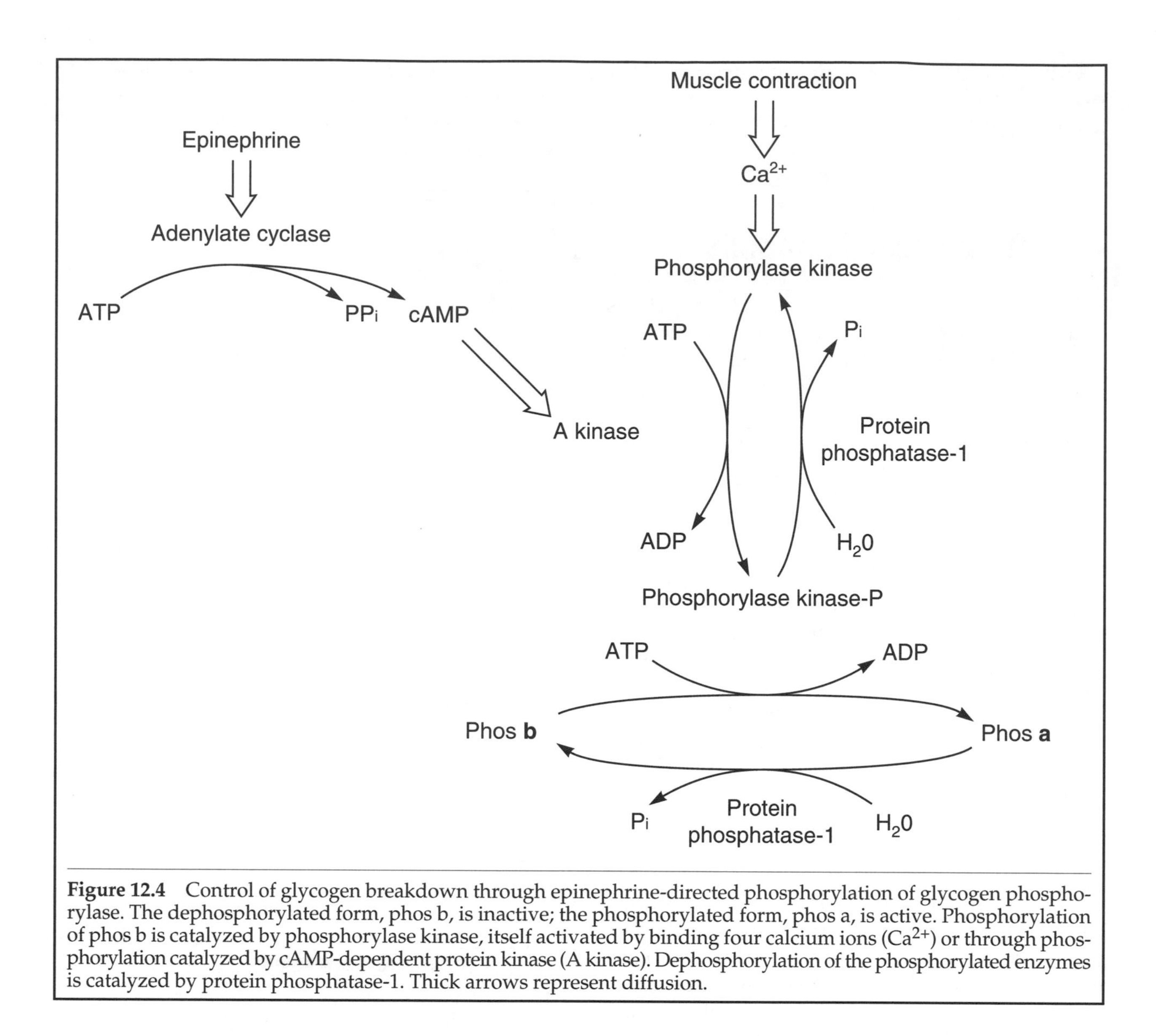

Figure 12.4 Control of glycogen breakdown through epinephrine-directed phosphorylation of glycogen phosphorylase. The dephosphorylated form, phos b, is inactive; the phosphorylated form, phos a, is active. Phosphorylation of phos b is catalyzed by phosphorylase kinase, itself activated by binding four calcium ions (Ca^{2+}) or through phosphorylation catalyzed by cAMP-dependent protein kinase (A kinase). Dephosphorylation of the phosphorylated enzymes is catalyzed by protein phosphatase-1. Thick arrows represent diffusion.

b have fairly high K_m values for one of their substrates, Pi. When muscle contracts, phosphocreatine (or creatine phosphate) decreases, and the Pi concentration rises. Pi must thus increase in the cytosol before significant glycogen is broken down in muscle, even if the active phosphorylase a is present. In the test tube, binding AMP activates phosphorylase b by operating as a positive allosteric effector. The AMP concentration increases in vigorously contracting muscle, but how important AMP activation of phosphorylase b is from a physiological perspective is not well understood. The rate of glycogenolysis in muscle also depends on the amount of glycogen in a muscle fiber. For the same contraction conditions, the higher the glycogen the more rapid the rate of glycogen breakdown. As glycogen decreases during exercise, the protein-glycogen particle releases phosphorylase. As a result, the phosphorylase is less able to be activated by phosphorylase kinase, becoming more available to inactivation through dephosphorylation by protein phosphatase-1.

Regulation of Glycogenolysis in Liver

Regulation of glycogenolysis in liver is similar to that in muscle, although glucagon stimulates adenylate cyclase, generating cAMP, in liver, whereas epinephrine does so in muscle. A nerve impulse in liver also releases calcium ions from intracellular stores thus activating liver phosphorylase.

Regulation of Glycogenesis in Muscle

Like glycogen phosphorylase, glycogen synthase exists in two forms. Glycogen synthase I (GS I) is the nonphosphorylated form which is normally active. The I stands for independent, that is, GS I is independent of the glucose 6-P concentration. Glycogen synthase D (GS D) is the phosphorylated form that is normally inactive but can become active if the glucose 6-P concentration increases. The D means that GS D is dependent on the glucose 6-P concentration; that is, it is allosterically activated by glucose 6-P. Figure 12.5 illustrates this control of glycogen synthase.

Phosphorylation inactivates the active form of glycogen synthase (GS I) either by phosphorylase kinase or by cAMP-dependent protein kinase (protein kinase A). Phosphorylase kinase, which also phosphorylates phosphorylase b, phosphorylates GS I. The phosphorylase kinase is activated by phosphorylation or by binding Ca^{2+}. Phosphorylation of GS I forms GS D, which is normally inactive unless glucose 6-P concentration increases. Conversion of inactive GS D to active GS I occurs by dephosphorylation, catalyzed by protein phosphatase-1. Insulin enhances the effect of the phosphatase, stimulating the formation of GS I to help make glycogen. Glycogen can also inhibit active GS I so that the liver and muscles do not store too much glycogen.

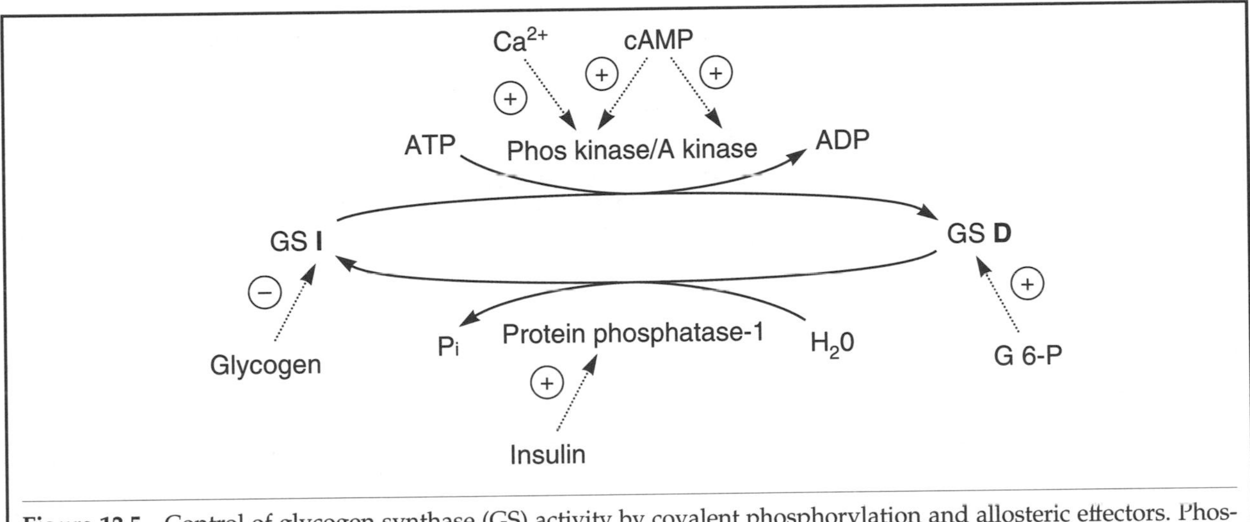

Figure 12.5 Control of glycogen synthase (GS) activity by covalent phosphorylation and allosteric effectors. Phosphorylation of the active form, GS I, generates the less active GS D form. Allosteric effectors and the enzymes they influence are shown with dotted arrows. Positive and negative signs indicate increase or decrease in activity, respectively.

Summary

Glycogen, the storage form of glucose in liver and muscle tissue, is synthesized from glucose taken up from the blood, in a process that is greatly accelerated following a carbohydrate-rich meal. Liver glycogen is a reservoir of glucose for the blood. In muscle, the glycogen is ready to be fed into the glycolytic pathway when the muscle fiber becomes active. Since glycogen synthesis and glycogenolysis are opposing processes taking place in the cytosol of cells, one process must be inactive when the other is active to avoid a futile cycle. Two regulatory enzymes, glycogen phosphorylase and glycogen synthase, are thus affected in opposite ways by the same signal. This is accomplished by attaching a phosphate group to serine residues in both enzymes. Phosphorylation of the synthase makes it less active; phosphorylation of phosphorylase makes it more active. In addition to covalent modification, the activity of both synthase and phosphorylase are influenced by the binding of regulatory molecules. Following a meal, when blood insulin concentration is increased, the synthase is active and the phosphorylase inactive in both liver and muscle. Between meals, liver phosphorylase is activated by a rise in glucagon and a fall in insulin so that the glycogen is slowly broken down to glucose. During exercise, when epinephrine (adrenaline) concentration is increased, muscle phosphorylase is active and synthase inhibited. The tight regulation of glycogen metabolism emphasizes the importance of controlling body carbohydrate stores.

CHAPTER 13

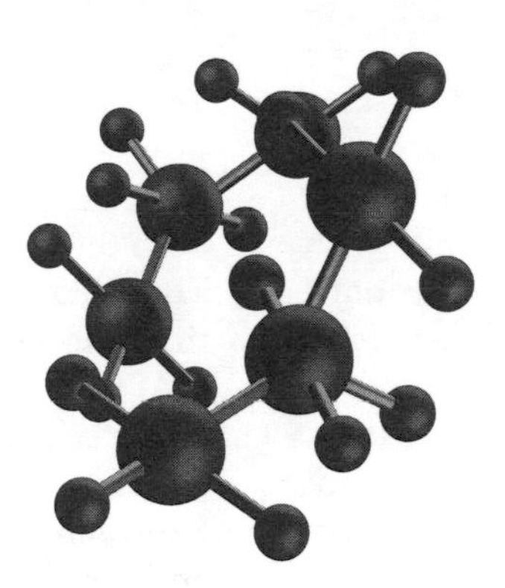

Amino Acid Metabolism

Humans typically ingest about 10 to 15% of their calories in the form of dietary protein. During digestion, protein is broken down to free amino acids, which are absorbed into the blood. As adults, the protein content of our bodies is remarkably constant, and so we might expect that we would oxidize amino acids to the tune of about 10 to 15% of our daily energy expenditure because no stores of amino acids exist. Those not used in protein synthesis or converted to other substances (e.g., heme groups; hormones, such as serotonin, adrenaline, and noradrenaline; synthesizing nucleotides; etc.) are simply used as fuels. They first lose their amino group(s) or other N atoms, and the resulting carbon skeleton can be oxidized directly, used to make glucose (gluconeogenesis) or converted into fat for storage.

Degradation of Amino Acids

Amino acids undergo constant oxidative degradation, during the following metabolic circumstances:

1. During normal synthesis and degradation of proteins in the body, amino acids released during the constant breakdown process may not be immediately reused in synthesis. Since we cannot store amino acids, they will be degraded.
2. When we ingest more amino acids than our bodies can use to make proteins or convert to other substances, these amino acids are degraded. In North America, most people eat more protein than they need.
3. During starvation, fasting, dieting, or uncontrolled diabetes mellitus, when carbohydrates are either not available or properly used, amino acid catabolism accelerates.

Catabolism of individual amino acids has two major stages. First, the amino acids must lose their nitrogen atoms because we cannot obtain usable energy from nitrogen groups. Second, the resulting carbon skeletons are fed into specific energy-yielding pathways to retrieve their chemical energy.

The liver removes the amino groups from most amino acids, although skeletal muscle removes most amino groups for the branched-chain amino acids (BCAAs). Exercise increases the catabolism of BCAAs by muscle. This fact explains the use of BCAAs by a number of athletes.

Transamination and Deamination

We have already seen transamination in the catabolism of BCAAs. In this reaction, vitamin B_6-containing transaminase enzymes (aminotransferases) transfer the amino group of an amino acid to α-ketoglutarate, making glutamate and a new keto acid (see Figure 13.1). Notice that the ketone group in α-ketoglutarate is next to the carboxylate. Nearly all transaminase reactions involve amino transfer between an amino acid and α-ketoglutarate, forming a new α-keto acid and glutamate. Additional enzymes are specific for amino acids other than glutamate. All transamination reactions are freely reversible with equilibrium constants of about one and standard free energy change values near zero. Their net direction depends on the relative concentrations of the four reactants. The net effect is to transfer amino groups from a variety of amino acids onto one amino acid, typically glutamate. Figure 13.1 summarizes the reactions for muscle, which preferentially transaminates BCAA, and liver, which handles most of the other amino acids. Figure 11.3 also shows, in skeletal muscle, the transfer of amino groups on the BCAAs to form alanine by two different transaminase enzymes.

The body rids itself of amino groups by forming urea. Therefore, the amino group on glutamate (which comes from amino acids) must be transferred to the liver (if not already there) and into the urea molecule. Nitrogen, from amino groups in the liver in the form of glutamate, can be released as ammonia (principally the ammonium ion NH_4^+) in the glutamate dehydrogenase reaction:

$$\text{glutamate} + H_2O + NAD^+ \xleftrightarrow{\text{glutamate dehydrogenase}} \alpha\text{-ketoglurate} + NADH + H^+ + NH_4^+$$

The freely reversible glutamate dehydrogenase reaction takes place only in the mitochondrial matrix, whereas most aminotransferase enzymes exist in both the mitochondrial matrix and the cell cytosol. The glutamate dehydrogenase reaction, in the direction shown, is a deamination reaction. Because NADH is also formed, it is also an oxidative deamination reaction. The glutamate dehydrogenase reaction removes nitrogen from glutamate to make urea. Exercising muscle releases considerable amounts of ammonia (as the NH_4^+ ion) and it has been determined that the glutamate dehydrogenase reaction in muscle is a major source of this ammonia. In fact, the production of ammonium ion and its release from muscle is proportional to the intensity of the exercise.

The content of the amino acid glutamine is high in a variety of cells. It is particularly abundant in skeletal muscle, out of proportion to the amount of glutamine in muscle proteins. Glutamine can be synthesized from glutamate according to the following equation. Notice that the additional nitrogen in glutamine is an amide of the side chain carboxylate (see Figure 13.2).

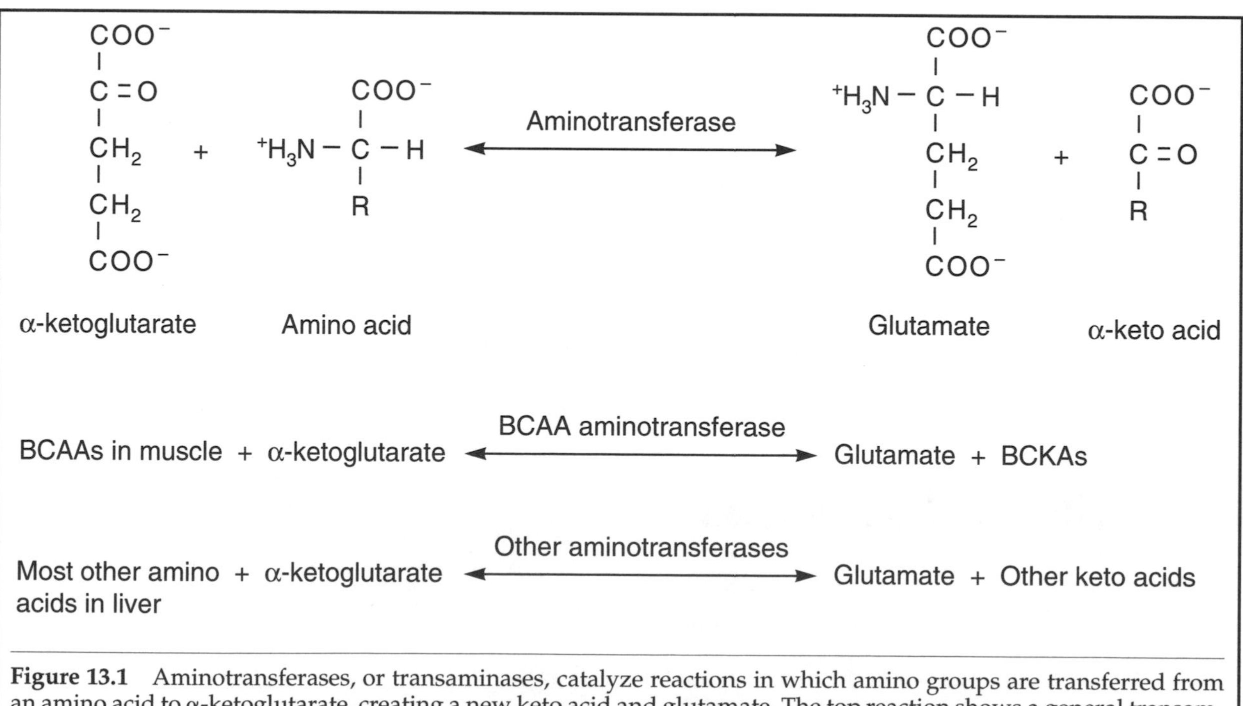

Figure 13.1 Aminotransferases, or transaminases, catalyze reactions in which amino groups are transferred from an amino acid to α-ketoglutarate, creating a new keto acid and glutamate. The top reaction shows a general transamination reaction using chemical structures. R is the side chain of the general amino acid. The middle equation shows transaminations that typically occur in muscle using the branched-chain amino acids (BCAAs) and generating branched-chain keto acids (BCKAs), whereas the bottom equation shows the general transamination reaction of other amino acids that usually take place in the liver.

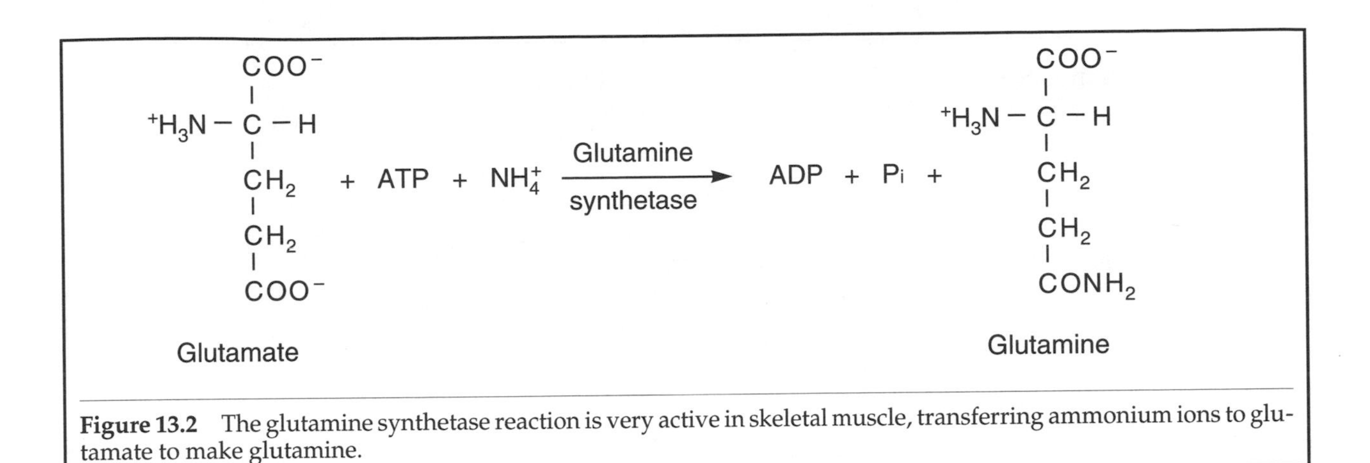

Figure 13.2 The glutamine synthetase reaction is very active in skeletal muscle, transferring ammonium ions to glutamate to make glutamine.

The last reaction to consider is the removal of the amino group from the side chain of glutamine, which is the reverse of the glutamine synthetase reaction. Glutaminase, found mainly in liver mitochondria, catalyzes this deamination reaction:

$$\text{glutamine} + H_2O \xrightarrow{\text{glutaminase}} \text{glutamate} + NH_4^+$$

As shown, skeletal muscle releases ammonia as a result of the deamination of glutamate. Skeletal muscle also releases alanine from a transamination reaction catalyzed by alanine aminotransferase, in which the amino group on glutamate is passed to pyruvate to make alanine (see Figures 11.3 and 11.4). The release of alanine from muscle accelerates during exercise, due in part to an increase in pyruvate concentration from glycolysis. The alanine formed in muscle is released and travels in the blood to the liver. Here, in another transamination reaction, the amino group of alanine is added to α-ketoglutarate to make glutamate, whereas the carbon skeleton remaining from alanine becomes pyruvate. This pyruvate is now a precursor for making glucose in the liver, and it can subsequently travel back to the same muscle to be used for glycolysis. This glucose-alanine cycle is very active during exercise. Skeletal muscle also releases glutamine, especially during exercise. Amination of glutamate, using the glutamine synthetase reaction (shown in Figure 13.2), must be the major source of glutamine, because muscle proteins are not rich in the amino acid glutamine.

The Urea Cycle

Ammonia is quite toxic, especially to the brain. However, two safe forms of ammonia exist: the amino group of glutamate and the side chain amide nitrogen in glutamine. Although we can temporarily store ammonia in these innocuous forms, we must eliminate the nitrogen we cannot use. Animals convert the nitrogen to urea, whereas birds and reptiles eliminate amino groups by converting nitrogen to uric acid. Urea, whose structure follows, is a simple molecule formed in the liver and excreted from the kidney when urine is formed.

$$H_2N\text{-}\overset{\overset{\displaystyle O}{\|}}{C}\text{-}NH_2$$

The two amino groups in urea allow the body to rid itself of nitrogen. The carbonyl group comes from carbon dioxide. The nitrogens come from the ammonium ion and from the amino group of aspartate. Figure 13.3 summarizes the path taken by nitrogen from amino acids in the body.

Muscle releases alanine, which contains much of the muscle nitrogen. In the liver, the amino group on alanine is transferred to glutamate, catalyzed by alanine aminotransferase. The glutamine released from muscle results from amination of glutamate in the glutamine synthetase reaction. The liver is the major site for nitrogen removal from most of the amino acids (except the BCAAs). The amino groups end up on glutamate. Muscle also releases ammonium ions, especially during exercise. Muscle glutamate is a major source of ammonium ion by the way of the glutamate dehydrogenase reaction.

Urea synthesis requires the ammonium ion that comes from glutamine via the glutaminase reaction or from glutamate via the glutamate dehydrogenase reaction. Aspartate, which provides the other urea nitrogen, comes from a transamination reaction in which oxaloacetate is transaminated from glutamate to make aspartate and α-ketoglutarate. The enzyme aspartate aminotransferase catalyzes the following reaction.

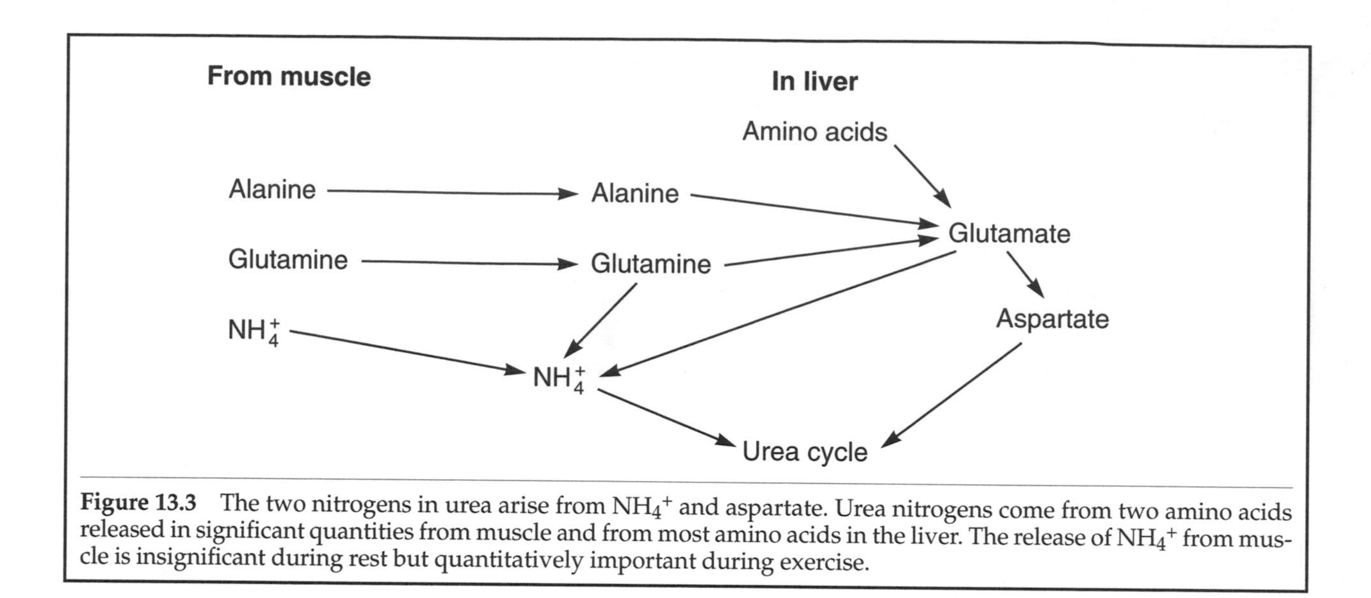

Figure 13.3 The two nitrogens in urea arise from NH_4^+ and aspartate. Urea nitrogens come from two amino acids released in significant quantities from muscle and from most amino acids in the liver. The release of NH_4^+ from muscle is insignificant during rest but quantitatively important during exercise.

$$\text{glutamate} + \text{oxaloacetate} \xleftrightarrow[]{\text{aspartate aminotransferase}} \alpha\text{-ketoglutarate} + \text{aspartate}$$

The urea cycle consists of four enzymatic steps that take place in the liver cells, both in the mitochondrial matrix and cytosol. Figure 13.4 illustrates the first reaction before the actual urea cycle begins. This first step occurs in the mitochondrial matrix and involves the formation of carbamoyl phosphate, a molecule containing nitrogen from the ammonium ion and a carbonyl group from the bicarbonate ion.

The carbamoyl phosphate synthetase reaction controls the rate of the urea cycle, which is shown in Figure 13.5. The carbamoyl phosphate enters the urea cycle in the matrix, joining with ornithine to form citrulline, in a reaction catalyzed by ornithine transcarbamoylase. The citrulline exits the mitochondrial matrix. The second nitrogen, in the form of the aspartate amino group, enters the cycle when aspartate combines with citrulline to form argininosuccinate. This step involves ATP hydrolysis and is catalyzed by argininosuccinate synthetase. The argininosuccinate is then cleaved to arginine and fumarate in a reaction catalyzed by argininosuccinate lyase. In the final step, arginine is cleaved by arginase to yield ornithine and urea. The ornithine is now regenerated and enters the mitochondrion to combine again with carbamoyl phosphate.

Urea is extremely water soluble. Thus it leaves the liver via the blood and enters the kidneys, where it is filtered out. The urea cycle is metabolically expensive because the ATP cost to make one urea molecule is four. Two are used to make carbamoyl phosphate. Can you think where the other two ATP come from? (Hint: How many phosphate groups are removed from ATP during the condensation of citrulline and aspartate?)

Fate of Amino Acid Carbon Skeletons

After the amino groups are removed from the amino acids, carbon skeletons remain, in many cases in α-keto acids, such as pyruvate, oxaloacetate, or α-ketoglutarate. These carbon skeletons have a variety of fates, such as gluconeogenesis, because 18 of the 20 amino acids can be a source of glucose. The carbon skeletons can also be used for immediate oxidation because they form citric acid cycle intermediates or acetyl CoA. Finally, all are potential sources of carbon to make new fatty acids because all are also potential sources of acetyl CoA. Figure 13.6 illustrates what the carbon skeletons of the amino acids have in common with the citric acid cycle or substances directly related to this cycle. The amino acids shown can generate the citric acid cycle intermediates or related molecules by simple removal of their amino groups via transamination (alanine, glutamate, aspartate) or through a number of steps not shown.

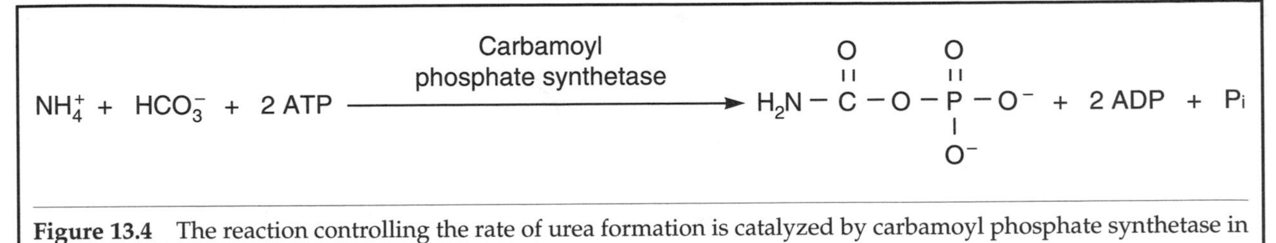

Figure 13.4 The reaction controlling the rate of urea formation is catalyzed by carbamoyl phosphate synthetase in the matrix of liver mitochondria.

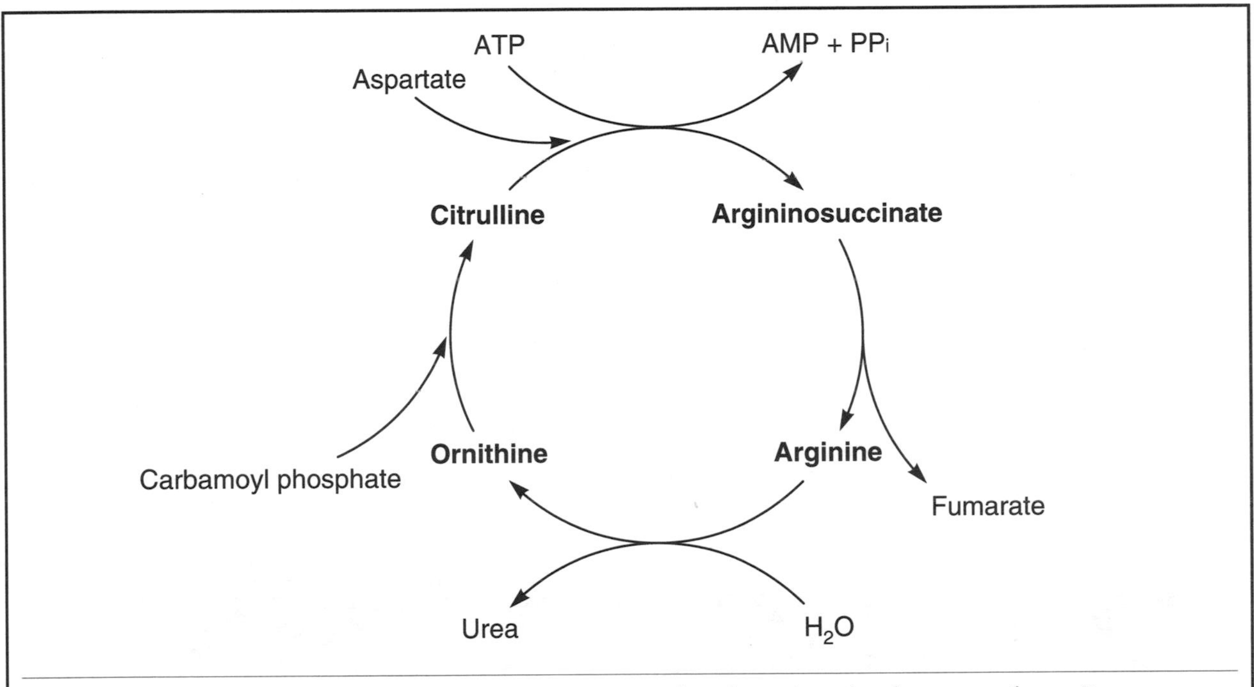

Figure 13.5 The urea cycle is a sequence of four enzyme-catalyzed reactions that forms urea from nitrogen on carbamoyl phosphate and aspartate. There is no net consumption or formation of ornithine or other intermediates in the cycle, shown in bolded letters.

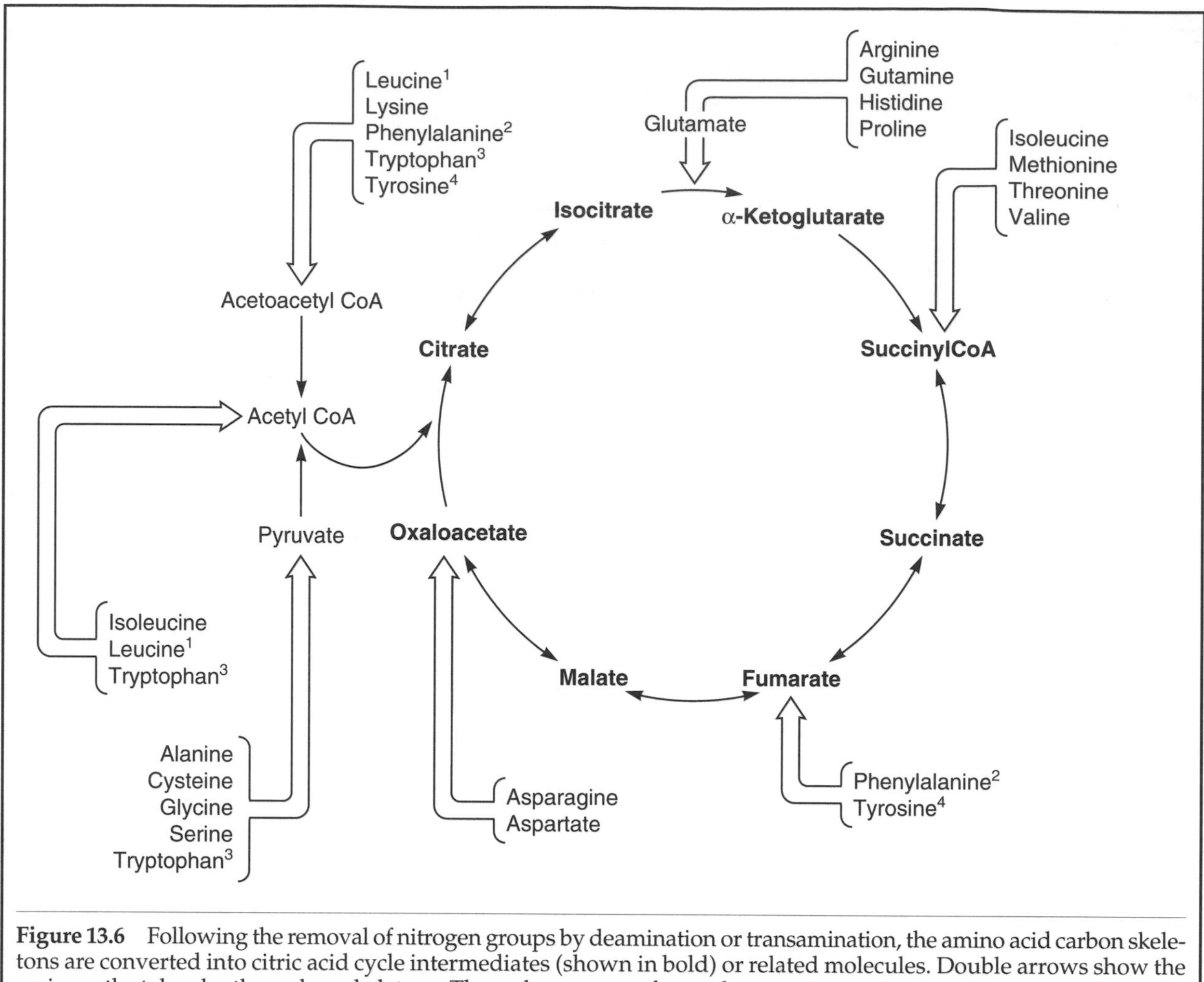

Figure 13.6 Following the removal of nitrogen groups by deamination or transamination, the amino acid carbon skeletons are converted into citric acid cycle intermediates (shown in bold) or related molecules. Double arrows show the major paths taken by the carbon skeletons. The carbon atoms of some large amino acids give rise to two or three different molecules, identified with superscripts.

Summary

The content of protein in the adult body remains remarkably constant. Because most adults take in about 10 to 15% of their dietary energy in the form of protein, an equivalent amount of amino acids must be lost each day. The body cannot store excess amino acids. Those not needed to make protein lose their nitrogen groups, and the carbon skeletons are used to make glucose (gluconeogenesis) or converted to acetyl CoA, which can feed into the citric acid cycle or be used to make fatty acids. The first step in disposing of excess amino acids is to remove the amino groups, which are primarily transferred to alpha-ketoglutarate by transaminase (aminotransferase) enzymes to produce glutamate and keto acids. This process occurs in the liver for most amino acids, but the branched-chain amino acids (leucine, isoleucine, and valine) lose their amino groups primarily in muscle through the action of branched-chain amino acid transaminase. Skeletal muscle contains a high concentration of glutamine, synthesized by glutamine synthetase from glutamate and ammonium ions. During exercise, muscle releases both alanine, ammonium ions, and glutamine at accelerated rates. Amino groups on amino acids are removed from the body in the form of urea, a molecule synthesized in liver using the urea cycle and excreted in the urine. This simple cycle eliminates amino groups that cannot be oxidized.

APPENDIX

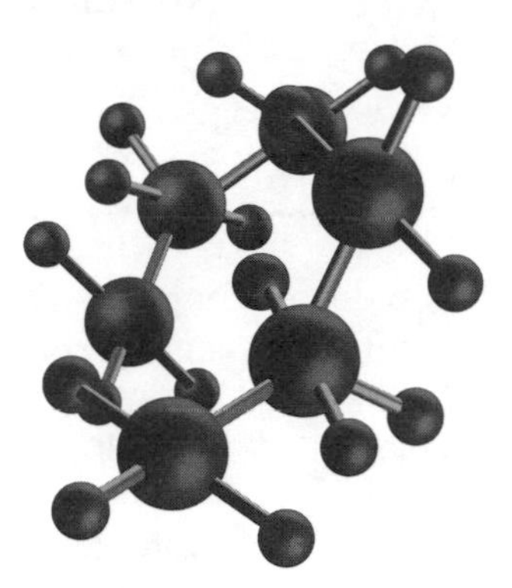

Review of Basic Chemistry

The following summarizes basic chemical concepts to aid in your understanding of biochemistry.

Ionic and Covalent Compounds

Ionic compounds form between metals and nonmetals when positively and negatively charged ions attract. The metals lose one or more electrons to nonmetals; thus the metal becomes a positively charged ion (cation), and the nonmetal becomes a negatively charged ion (anion). When dissolved in water, ionic compounds lose their solid crystal structure, and the individual ions are dispersed. For example, when solid calcium chloride ($CaCl_2$) is dissolved in water, a solution of Ca^{2+} and Cl^- ions is formed in which Cl^- are twice as abundant as Ca^{2+}.

Covalent compounds form when neighboring atoms share electrons. A single covalent bond, denoted by a dash, results when a pair of electrons is shared; a double bond, the sharing of two pairs of electrons, is indicated by two dashes. Most compounds in living organisms contain carbon with four covalent bonds, that is, four single bonds, two single and one double bond, or two double bonds. Nitrogen has three single bonds or one double and one single bond. Oxygen has two single bonds or a double bond, while hydrogen has only a single bond.

Not all covalent bonds involve equal sharing of electrons between adjacent atoms. Elements like nitrogen and oxygen tend to attract the paired electrons closer to them and farther away from the neighboring atom. Thus oxygen and nitrogen will be somewhat negatively charged and the neighboring atom somewhat positively charged. The covalent bond is said to be polar as are molecules containing polar bonds. Water has two polar bonds between the oxygen and each hydrogen, making it a polar compound. In solution, cations will be closer to the oxygen end of a water molecule, whereas anions will be closer to the hydrogen atoms.

Hydrogen bonds represent a weak electrostatic attraction between a covalently bonded hydrogen (H) atom and a covalently bonded electronegative atom (such as oxygen or nitrogen) on the same or a different molecule. These are weak bonds, about 1/25th as strong as a single covalent bond, but they are extremely important. The fact that water has a high melting point (O°C) and boiling point (100°C) and requires a large amount of energy to be vaporized despite its small size is due to extensive hydrogen bonding between adjacent water molecules.

The Mole Concept

A mole is 6.02×10^{23} particles of anything; this is known as Avogadro's number. The particles can be atoms, molecules, or ions. The atomic weight of carbon is 12; thus a mole of carbon (6.02×10^{23} atoms) weighs 12 g. The molecular weight of carbon dioxide (CO_2) is 44, so a mole of CO_2 (6.02×10^{23} molecules) weighs 44 g. The formula weight of calcium chloride ($CaCl_2$) is 111. A mole of $CaCl_2$ will weigh 111 g and in solution will contain one mole of Ca^{2+} ions and two moles of Cl^- ions.

Molarity (M) expresses the concentration of, or number of moles of, solute in a liter of solution. The molecular weight of glucose is 180. Thus a one molar solution of glucose will contain 180 g of glucose per liter of glucose solution, expressed as 1 M or 1 mole/L. A millimole (mM) contains one millimole per liter. The molecular weight of lactic acid is 90. If the blood lactic acid concentration is 10 mM, there are 900 mg of lactic acid per liter of blood. A micromole (μM) contains one millionth of a mole.

Acids and Bases

In earlier chemistry courses you may have learned that an acid solution was one in which the concentration of hydrogen ions ($[H^+]$) is greater than the concentration of hydroxide ions ($[OH^-]$). In a basic solution the concentration of hydrogen ions is less than the concentration of hydroxide ions ($[H^+] < [OH^-]$).

Dissociation of Water

Water is vital because (a) it forms the solvent phase of virtually all cells and body spaces, (b) it dissolves a range of ionic compounds and large and small polar molecules, and (c) the chemical reactions of life take place in a water medium. Water molecules have a limited tendency to dissociate or form ions. The following equation shows the dissociation or ionization of water:

$$H_2O \longleftrightarrow H^+ + OH^-$$

H^+ symbolizes the hydrogen ion; it is a proton. The OH^- symbolizes the hydroxide ion. The use of a double-headed arrow means that the dissociation of water does not completely generate hydrogen and hydroxide ions, but reaches an equilibrium with both reactants (H^+ and OH^-) and products (undissociated water) present.

The ionization of water occurs rapidly, but at 25°C and with pure water the concentration of hydrogen ions equals the concentration of hydroxide ions.

$$[H^+] = [OH^-] = 1 \times 10^{-7} \text{ moles} \times L^{-1}$$

Note that [ ] means concentration of. In pure water, the concentration of un-ionized (undissociated) water is

$$55.6 \text{ moles} \times L^{-1}.$$

Whenever we have an incomplete reaction, in which an equilibrium is reached, we can set up an equilibrium expression. For the ionization (dissociation) of water this expression is the equilibrium equation.

The equilibrium constant for the ionization of water (K) is expressed in the equation

$$[H^+]\,[OH^-] \,/\, [H_2O] = K.$$

As with any equilibrium expression, the multiplication product of the concentrations of the products at equilibrium is divided by the equilibrium concentration of the reactant(s). If we take the equilibrium equation and cross multiply, we get

$$[H^+]\,[OH^-] = K[H_2O] = K_w.$$

The term K_w, which is equivalent to $K[H_2O]$, is the ion product constant for water. Since at equilibrium in water,

$$[H^+] = [OH^-] = 1 \times 10^{-7} \text{ moles} \,/\, L,$$

the K_w will equal 1×10^{-14} at 25°C.

Concept of pH

The symbol pH expresses the acidity or alkalinity of a solution. It is the negative logarithm of the hydrogen ion (proton) concentration, written as

$$pH = -\log [H^+].$$

The concentration of protons is

$$1 \times 10^{-7} \text{ moles} \times L^{-1}.$$

Therefore the pH is equal to ($-\log 1 \times 10^{-7}$). This equation could be written as

$$-(\log 1 + \log 10^{-7})$$

which is equal to $-(0-7)$. Thus the pH of pure water is seven. If we add hydrochloric acid (HCl) to water at a concentration of 0.01M, the pH can be determined as follows. HCl is a strong acid that completely dissociates so

$$[H^+] = [Cl^-] = 0.01 \text{ or } 10^{-2} \text{ moles} \,/\, L^{-1}.$$

The pH will be two.

The pH is a logarithmic scale with zero being the low end or most acidic and 14 being the most basic. Therefore, if the concentration of protons is increased by a factor of 10, the pH will decrease by one.

Bronsted-Lowry Definition

An acid is a proton (H^+ or hydrogen ion) donor, and a base is a proton acceptor. We distinguish strong and weak acids by their tendency to donate a proton. Strong acids, such as HCl (hydrochloric acid), H_2SO_4 (sulfuric acid), $HClO_4$ (perchloric acid), and H_3PO_4 (phosphoric acid), completely give up their protons. For example, we can write the dissociation of HCl as

$$HCl \longrightarrow H^+ + Cl^-.$$

On the other hand, moderate to weak acids, such as HSO_4^- (hydrogen sulfate ion), $H_2PO_4^-$ (dihydrogen phosphate ion), and CH_3COOH (acetic acid), do not dissociate completely. For the dissociation of the phosphate ion we can write

$$H_2PO_4^- \longleftrightarrow HPO_4^{2-} + H^+.$$

$H_2PO_4^-$ is the acid and HPO_4^{2-} its conjugate base. The equilibrium equation to demonstrate this dissociation is

$$[HPO_4^{2-}][H^+]/[H_2PO_4^-] = k_a$$

where k_a is the acid dissociation constant—like the equilibrium constant k, with the subscript indicating acids. For the dissociation of the phosphate ion, the k_a is 1.6×10^{-7}. The stronger the acid, the greater the dissociation, the larger the numerator, the smaller the amount of undissociated acid, and thus the larger the value of k_a.

The ammonium ion (NH_4^+) dissociates to form NH_3 (ammonia) and a proton.

$$NH_4^+ \longleftrightarrow NH_3 + H^+$$

The k_a will be described by

$$[NH_3][H^+]/[NH_4^+] = 5.6 + 10^{-10}.$$

Because its k_a is larger, we know that the dihydrogen phosphate ion ($H_2PO_4^{2-}$) is a stronger acid than the ammonium ion. However, ammonia (NH_3) is a stronger base than the monohydrogen phosphate ion (HPO_4^{2-}).

To avoid the use of exponents when using k_a values, as when describing [H^+], we take the negative log of k_a and get a pk_a. That is,

$$pk_a = -\log k_a.$$

Because we use a negative logarithm, the smaller the pk_a the stronger the acid.

Buffers

The proper functioning of our bodies requires a constant or near constant pH of fluids. We thus need substances to reduce the effect of acids and bases. These substances are buffers. A buffered solution contains either a weak acid and its conjugate base or a weak base and its conjugate acid. Carbonic acid (H_2CO_3) and the bicarbonate ion, its conjugate base (HCO_3^-), represent an important buffer system. Dihydrogen phosphate ($H_2PO_4^-$) and monohydrogen phosphate (HPO_4^{2-}), its conjugate base, are other important buffers.

The Henderson-Hasselbalch equation shows how to determine the pH of a buffer solution if the pk_a of the acid and the concentration of undissociated acid and its conjugate base are known.

$$pH = pk_a + \log [\text{conjugate base}]/[\text{acid}]$$

Organic Chemistry

Most chemical substances in the body are organic molecules, which contain carbon and hydrogen. In living cells, most organic molecules also contain oxygen, whereas amino acids contain nitrogen. The chemical bonds between carbon atoms in most organic molecules are single bonds. Carbon-to-carbon double bonds are indicated by (=). Organic functional groups attached to the basic hydrocarbon molecules are responsible for many of their physical and chemical properties. Figure A.1 illustrates common functional groups encountered in biochemistry.

Alcohols, containing the OH or hydroxy group, are common in many *in vivo* molecules. The carbonyl group is found in aldehydes, ketones, esters, amides, and carboxylic acids. Alcohols are called primary, secondary, or tertiary depending on whether the carbon to which the hydroxy group is attached is itself bonded to one, two, or three other carbon atoms, respectively. Figure A.1 shows several different ways of indicating the structure of certain functional groups. For example, the double bond between the carbonyl carbon and oxygen can be shown in detail or implied in aldehydes, carboxylic acids, and ketones. Phosphoric acid may be drawn showing all covalent bonds or in shorthand to suggest their presence. Aromatic compounds are organic molecules containing a benzene ring structure.

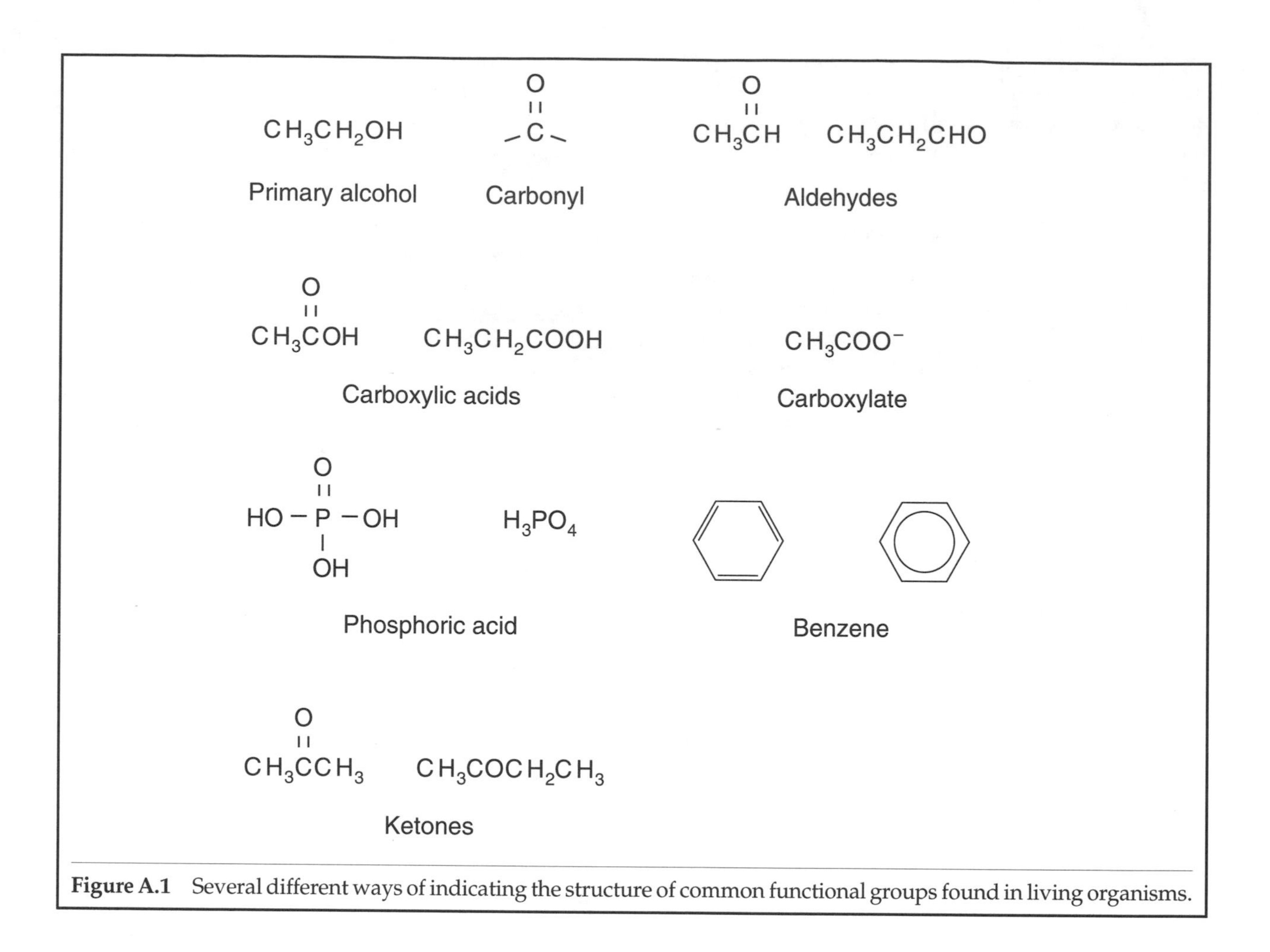

Figure A.1 Several different ways of indicating the structure of common functional groups found in living organisms.

Figure A.2 indicates the relationships between functional groups. Oxidation of the alcohol group to a carbonyl group is an important metabolic reaction. Oxidation of a primary alcohol produces an aldehyde; oxidation of a secondary alcohol produces a ketone, whereas tertiary alcohols cannot be oxidized. Because of its ready dissociation, the carboxyl group *in vivo* is in its conjugate base form, known as the carboxylate group. The carboxyl group (actually carboxylate *in vivo*) reacts with an alcohol group to produce an ester, with an amine to produce an amide, and with another carboxyl group to produce an anhydride. Since the phosphate ion is derived from phosphoric acid, it combines with an alcohol to form a phosphate ester and with another phosphate group to form a phosphate anhydride.

Molecules also exist that have two or more functional groups. For example, carboxylic acids with a ketone group are keto acids. An acid with a hydroxy (hydroxyl) group is a hydroxy acid. Finally, in some molecules a sulfur atom(s) replaces an oxygen. Molecules with SH at the end (corresponding to OH) are called thiols.

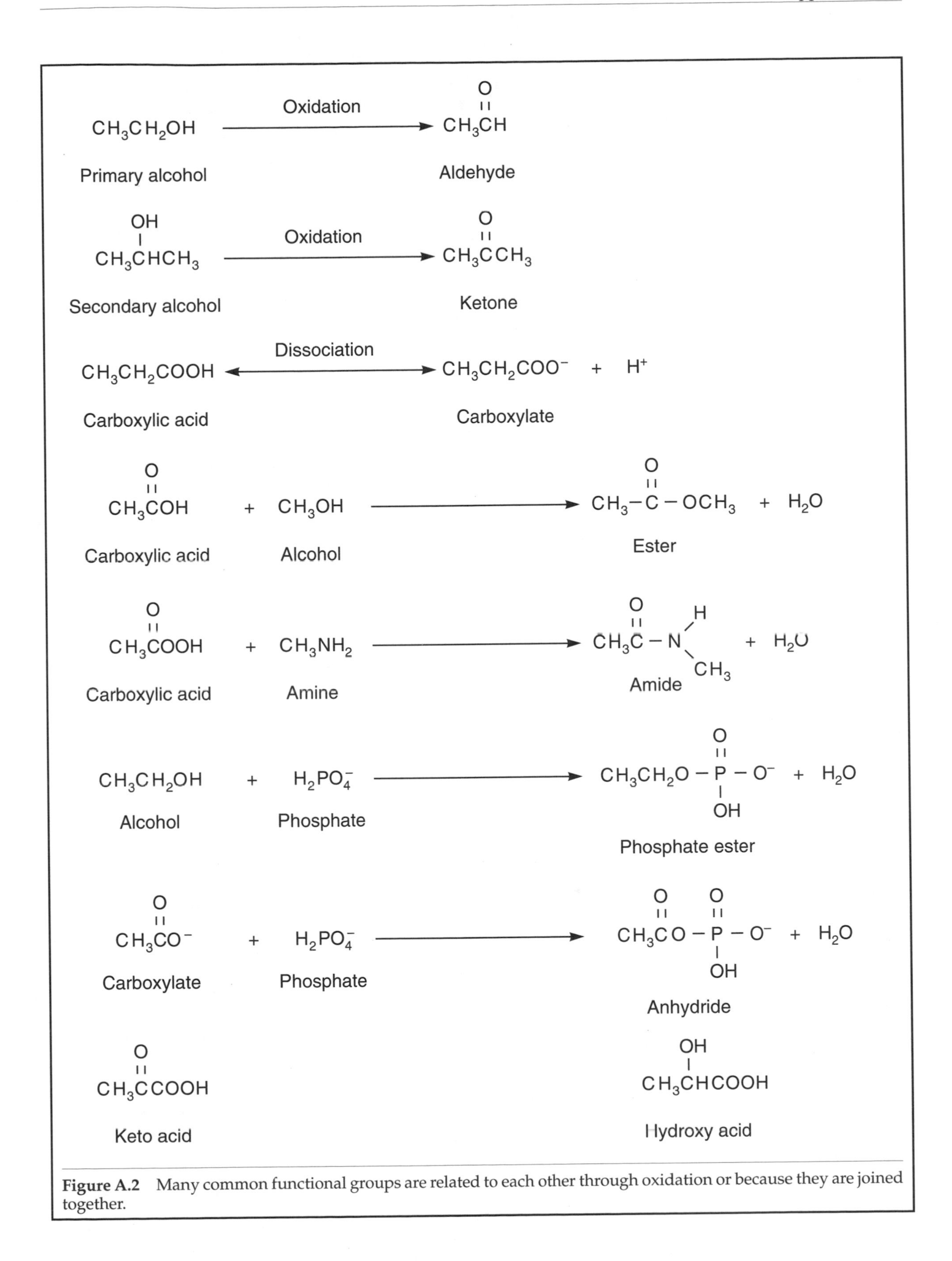

Figure A.2 Many common functional groups are related to each other through oxidation or because they are joined together.

BIBLIOGRAPHY

Brown, G.C. (1992). Control of respiration and ATP synthesis in mammalian mitochondria and cells. *Biochemical Journal*, **284**, 1–13.

Cohen, P. (1992). Signal integration at the level of protein kinases, protein phosphatases and their substrates. *Trends in Biochemical Sciences*, **17**, 408–413.

Hershey, J.W.B. (1991). Translation control in mammalian cells. *Annual Reviews of Biochemistry*, **60**, 717–755.

Krajewska, W.M. (1992). Regulation of transcription in eukaryotes by DNA-binding proteins. *International Journal of Biochemistry*, **24**, 1885–1898.

Lafontan, M., & Berlan, M. (1993). Fat cell adrenergic receptors and the control of white and brown fat cell function. *Journal of Lipid Research*, **34**, 1057–1091.

Lehninger, A.L., Nelson, D.L., & Cox, M.M. (1993). *Principles of Biochemistry*. New York: Worth.

Lewin, B. (1990). *Genes IV*. Cambridge: Oxford University Press.

McMahon, S.B., & Monroe, J.G. (1993). Role of primary response genes in generating cellular responses to growth factors. *FASEB Journal*, **6**, 2707–2715.

Pilkis, S.J., & Granner, D.K. (1992). Molecular physiology of the regulation of hepatic gluconeogenesis and glycolysis. *Annual Reviews of Physiology*, **54**, 885–909.

Reichel, R.R., & Jacob, S.T. (1993). Control of gene expression by lipophilic hormones. *FASEB Journal*, **7**, 427–436.

Rhoads, R.E. (1993). Regulation of eukaryotic protein synthesis by initiation factors. *Journal of Biological Chemistry*, **268**, 3017–3020.

Sachs, A.B. (1993). Messenger RNA degradation in eukaryotes. *Cell*, **74**, 413–421.

Index

L

M

N

O

P

R

S

T

U

V

Z

About the Author

Michael Houston completed his PhD in 1969 before joining the Department of Kinesiology at the University of Waterloo, where he has taught biochemistry to exercise science students for more than 25 years. In addition to biochemistry, Dr. Houston teaches a graduate course in molecular biology and muscle chemistry for exercise physiology students. One of his major teaching interests is to prepare exercise physiology students for the molecular developments at the forefront of the exercise physiology field.

Dr. Houston is currently a professor in the kinesiology department at the University of Waterloo. He is past president of the Canadian Society for Exercise Physiology and a member of the American College of Sports Medicine. He has published more than 80 articles in prominent scholarly journals on the integration of biochemical and physiological mechanisms in the exercising muscle.